Cleaner Transport Fuels for Cleaner Air in Central Asia and the Caucasus

Masami Kojima
Robert W. Bacon
Martin Fodor
Magda Lovei

The World Bank
Washington, D.C.

Cover design by Tomoko Hirata.

ISBN 0-8213-4783-7

Library of Congress Cataloging-in-Publication Data:
Cleaner transport fuels for cleaner air in Central Asia and the Caucasus / Masami Kojima
... [et al.].
      p. cm.
  Includes bibliographical references.
  ISBN 0-8213-4783-7
   1. Air quality management—Asia, Central. 2. Air quality management—Transcaucasia.
3. Motor fuels—Asia, Central. 4. Motor fuels—Transcaucasia. 5.
Automobiles—Motors—Exhaust gas—Environmental aspects—Asia, Central. 6.
Automobiles—Motors—Exhaust gas—Environmental aspects—Transcaucasia. I. Kojima,
Masami. II. World Bank.

HC420.3.Z9 A43 2000
363.738'7'09475—dc21
                                    00-043403

Table of Contents

ACKNOWLEDGMENTS

This paper describes the regional study Cleaner Transportation Fuels for Urban Air Quality Improvement in Central Asia and the Caucasus, undertaken jointly by the Energy and Environment Units of the Europe and Central Asia Region, the Environment Department, and the Oil, Gas and Chemicals Department of the World Bank Group, with support from the joint United Nations Development Programme (UNDP)/World Bank Energy Sector Management Assistance Programme (ESMAP) and the Canadian International Development Agency (CIDA). The financial assistance of the government of the United Kingdom through ESMAP is gratefully acknowledged.

The World Bank team includes Masami Kojima (task manager) and Robert Bacon of the Oil, Gas and Chemicals Department, and Magda Lovei (co-task manager) and Martin Fodor of the Environment Department. The staff of AEA Technology, Environment Canada, and SNC-Lavalin*Comcept Canada participated in the study and contributed to data collection and analysis in the eight participating countries, as did numerous local experts in the environment, energy, and transport sectors. These contributors included representatives from government agencies, refineries, academia, and nongovernmental organizations.

The authors are grateful to Hossein Razavi, director of the Europe and Central Asia Region Energy Department, and Kevin Cleaver, director of the Europe and Central Asia Region Environmentally and Socially Sustainable Development Department, for their guidance during the execution of this study. The authors also thank Todd Johnson and David Craig for their contributions as peer reviewers; Peter Thomson, Michele de Nevers, and Konrad von Ritter for their support and helpful comments; and Nancy Levine for editorial assistance.

Executive Summary

Urban air pollution is a matter of increasing concern in many of the newly independent states (NIS). Traffic is burgeoning in cities, but vehicle registration, inspection, and maintenance lag behind what is needed to support efforts to improve air quality. Poor fuel quality worsens the emissions problems. Although certain emissions, such as greenhouse gases, are of global concern, the greatest costs of air pollution at the local level are to human health. It is estimated that nearly 40,000 people die prematurely and about 100,000 people fall ill every year in big cities throughout the

NIS as a result of exposure to excessive air pollution (Hughes and Lovei 1999). The economic costs of these health impacts are estimated to reach as much as 5 percent of city incomes. Small and dispersed sources of pollution such as motor vehicles contribute more to human exposure than their share of total emissions loads. Of primary importance are fine particulate matter, implicated in respiratory diseases; and lead, which is injurious to children's mental development and is a persistent pollutant that accumulates in the environment.

The NIS countries have especially pressing problems with air pollution originating from transport because of their aging vehicle fleets and their lack of adequate infrastructure for fuel distribution, vehicle servicing, and inspection and maintenance. In many cities in Azerbaijan, Kazakhstan, and Uzbekistan, transport is said to be the main source of air pollution, accounting for a significant fraction of total urban pollution in some places. As a result of the economic downturn, total energy consumption in these countries declined in the 1990s, resulting in reduced emissions from the combustion of fossil fuels. Although in the short term the effect is to curb urban air pollution, the economic contraction has also decreased the resources available for pollution abatement and has led to slower renewal of the vehicle fleet, affecting the long-term prospects for urban air quality improvement. Over the next 10 to 15 years, the increase in the number of vehicles and growing vehicle use will lead to a steady rise in vehicle emissions of damaging pollutants if there are no changes in the current policies regarding fuel specifications and vehicle emissions standards.

To avoid this outcome, it is necessary to establish a framework for short- and medium-term measures for improving both fuel quality and vehicle emissions performance. Achieving just one of these goals would have a very limited impact. Furthermore, in order to determine the speed and the rigor with which policies should be implemented, it is important to know the nature and the magnitude of the pollution problem; hence, reliable air quality monitoring is needed. Measures to impose (and enforce) tighter air pollution standards have implications for domestic refineries, for the tax and tariff regime, and for traffic management. In other words, the problems are multisectoral, involving the energy, environment, and transport sectors. On the positive side is the availability of natural gas, a clean fuel, in several of the countries; the possibility of leapfrogging old technology; and the potential for transferring and applying lessons learned about air quality management in other countries.

Combating pollution is not simply a matter of phasing in the more stringent standards that prevail in North America and the European Union. Those standards are supported by a regulatory and physical infrastructure that is not always in place in countries making the transition from central planning to a market economy. Policies, regulatory measures, and investments should reflect the economic and environmental conditions and the institutional constraints of countries in Central Asia and the Caucasus region.

Policymakers in the region are recognizing the need to take steps against air pollution. An indication of their concern is their commitment to phase out lead from gasoline. This commitment builds on numerous national and regional environmental studies and strategies, as well as donor-supported environmental programs and awareness-building efforts. At the Fourth Environment for Europe Ministerial Conference, held in Århus, Denmark, in 1998, all of the members of the United Nations Economic Commission for Europe (UNECE) present adopted a regional strategy aimed at phasing out lead from gasoline by 2005.[1]

To support the countries' efforts, the World Bank has undertaken a regional study, Cleaner Transportation Fuels for Urban Air Quality Improvement in Central Asia and the Caucasus. Armenia, Azerbaijan, Georgia, Kazakhstan, the Kyrgyz Republic, Tajikistan, Turkmenistan, and Uzbekistan participated in the study, which built on the momentum of the Århus conference. The study recognizes that gasoline lead should be phased out within the broader context of fuel quality improvement, especially given that the refining sector in the region is in the midst of serious restructuring. Fuel quality requirements, in turn, should be closely linked to broader air quality management to ensure that cost-effectiveness is considered and that the efforts of various sectors and stakeholders are coordinated. Environmental problems, environmental and fuel regulations and practices, and air quality monitoring systems in the countries of the region have much in common. In addressing environmental and fuel quality issues, it is therefore possible to build on economies of scale, avoid duplications, allow the transfer of experience, and facilitate intraregional trade in petroleum products.

The study has found that the countries of Central Asia and the Caucasus face a number of inherited and new obstacles in dealing with air pollution from transport sources:
- An aging, often poorly maintained fleet
- Inadequate vehicle registration systems
- Inadequate inspection and repair facilities
- Inadequate air quality monitoring systems
- Inadequate fuel quality monitoring and enforcement

- Refineries that are in need of upgrading
- Fuel pricing that creates incentives to adulterate (or smuggle) gasoline.

The above findings, as well as detailed analyses of air quality, the current air quality monitoring system, characteristics of the vehicle fleet, projections of transport fuel consumption, and the downstream petroleum sector, led to several observations and recommendations:

Air quality monitoring. The current air quality monitoring system is often not operational. Some of the equipment cannot provide reliable data even for the purposes of the existing monitoring system and do not yield data that can be directly compared with international guidelines and standards such as the health-based air quality guidelines of the World Health Organization (WHO). The current system requires measurement of a large number of pollutants, but these do not necessarily include the pollutants relevant for decisionmaking in air quality management. The study recommends the following measures:
- Establish a system of continuous monitoring of some or all of the six pollutants termed "classical" pollutants by the WHO, to permit direct comparison with international guidelines and standards. Resources could be rationalized by reducing the number of pollutants monitored.
- Monitor fine particles rather than total suspended particles (TSP); it is fine particles that are implicated in adverse health effects.
- In cities where leaded gasoline is still in use, monitor lead for longer periods at a time rather than take monthly averages of discrete 20-minute measurements.
- Monitor ground-level ozone at more stations in cities where ozone levels are high.

Reduction of vehicular emissions. The consumption of diesel fuel is forecasted to climb rapidly as gasoline-powered heavy-duty vehicles now in use are retired. Emerging epidemiological evidence indicates that fine diesel particulate emissions are also carcinogens; thus this growth is likely to have a significant public health impact. It will be important to establish an effective inspection and maintenance (I/M) program. Although there is a shortage of emissions measurement equipment at present, purchasing more equipment in itself will not be effective unless steps are taken at the same time to identify high emitters and carry

---

[1] The exceptions are Armenia, Macedonia, the Russian Federation, Turkey, and Uzbekistan, which reserved their position on the final phaseout date and called for a delay until 2008.

out corrective steps. The most urgent needs, in order to have a good I/M system, are:

- Readily available service and repair facilities with good diagnostic equipment and qualified technicians
- A computerized, up-to-date vehicle registration system
- A means of screening vehicles—possibly through remote-sensing technology—so that limited enforcement resources can be targeted on "gross emitters."

Improvement of fuel characteristics. The two most significant changes expected to take place in the region in the coming decade are the replacement of heavy-duty gasoline vehicles by diesel vehicles and the phaseout of passenger cars using low-octane gasoline in favor of cars using high octane. The results of these changes will be a rapid growth of demand for diesel, a slow growth of demand for gasoline, and a rapid increase in the share of high-octane gasoline. The slow growth of gasoline consumption will make it possible to eliminate lead in gasoline and to begin to meet the octane requirements of the vehicle fleet at a small cost to consumers. Recommendations for gasoline and diesel quality specifications include the following maximum limits:

PROPOSED GASOLINE AND DIESEL SPECIFICATIONS, MAXIMUM LIMIT

Fuel	Grade	Parameter	2005	2015[a]
Gasoline	All	Lead	0.013 g/l	0.013 g/l
	All	Benzene	5 vol%	2 vol%
	All	Sulfur	No change	0.03 wt%
	A76/80	Aromatics	No limit	35 vol%
	A91/93/95	Aromatics	No limit	45 vol%
Diesel	Vehicle grade	Sulfur	0.2 wt%	0.05 wt%

Note: g/l, gram per liter; vol%, percentage by volume; wt%, percentage by weight

a. The timing and the compositional limits should be reassessed in a few years' time.

Specifically:

- Eliminate lead in gasoline by 2005.
- Limit sulfur in gasoline to a level compatible with the efficient operation of catalytic converters, by 2015.
- Phase in reductions in benzene and total aromatics by 2015 (or earlier), giving refineries time to adjust their operations away from reliance on these compounds.

- Reduce diesel sulfur by 2005 and implement by 2015 the same diesel sulfur specification as was introduced in the United States and the European Union in the 1990s.
- Where air pollution problems are severe, consider introducing these specifications sooner or adding such specifications as standards aimed at reducing summer ozone levels.
- Improve the system for monitoring and enforcing compliance with fuel standards.

Implementation. The principal challenge for refineries is to phase out lead, increase average octane, and control benzene and total aromatics. It is important to emphasize that measures to follow the above recommendations will have to be incorporated into the ongoing efforts in refinery restructuring and upgrading. The incremental costs of meeting changing market demand and adopting the proposed fuel quality improvements are estimated to be on the order of US$0.01 per liter of gasoline. The necessary up-front investment costs are typically between US$20 million and US$50 million, depending on the refinery. In countries with a state-owned domestic refining sector, the capital required to modernize the refineries is likely to be difficult to raise. Sector deregulation and the entry of the private sector are expected to be beneficial not only for the performance of the sector but also for its ability to provide cleaner fuels. The study's recommendations include the following:

- Consider the installation of isomerization units and the purchase of oxygenates in order to limit benzene and aromatics.
- Reevaluate the role of "mini" refineries in gasoline production.
- Include expected current and future fuel quality specifications in privatization bidding documents and contracts to ensure predictability of regulations.

Regional harmonization and cooperation. Concerted action within and among countries is important for optimal pollution control policies. It is recommended that countries:

- Consider adopting uniform fuel standards throughout the region to reduce incentives for smuggling, to facilitate control of fuel quality, and to benefit from greater intraregional trade.
- Explore the potential for the private sector, including nongovernmental organizations, to take over some monitoring and enforcement responsibilities, relieving the government of those tasks.
- Encourage close coordination among the ministries of environment, transport, and energy, as well as

with the police and other agencies, in implementing the air quality management strategy.

- Disseminate environmental information and encourage and promote public education and professional training in vehicle repair and maintenance, proper fuel use, and environmental health issues.

Capacity building and the role of donors. Improving urban air quality is an important part of enhancing the quality of urban life and protecting people's health. Building commitment and capacity to take necessary measures will require a concerted effort by many stakeholders, including civil society, government, and industry. The development community can play an important role in supporting countries in the region in their endeavor by:

- Transferring applicable experience in air quality management and monitoring, fuel quality regulation and control, and vehicle emission regulation and abatement programs
- Helping to establish a network for learning and building endogenous institutions and capacity
- Piloting air quality management mechanisms and solutions that are applicable to local conditions
- Assisting with the establishment of a predictable policy and regulatory framework that will help attract the private sector financing needed to implement the recommendations for the refining sector
- Facilitating the removal of institutional and market barriers to involvement of the private sector in areas such as fuel quality monitoring, air quality monitoring, and vehicle inspection and maintenance programs.

Background

Urban air pollution has become a matter of concern in Central and Eastern Europe (CEE) and the new independent states (NIS). Urban traffic is increasing, the vehicle fleet still uses low-quality fuel, and the inherited monitoring and enforcement systems are not adequate for dealing with the new challenges. The air monitoring procedures are not yet compatible with World Health Organization (WHO) recommendations, making it difficult to compare air quality in the region with international guidelines.

The greatest cost of air pollution is that to human health. International experience indicates that, typically, *lead* and *fine particulate matter* are the greatest health concerns in the urban environment. Phasing out lead may be the most urgent priority in dealing with urban environmental problems (see Box 1). In 1996 the United Nations Economic Commission for Europe (UNECE) established a Task Force to Phase Out Leaded Gasoline, with the participation of Western European countries, CEE and NIS countries in transition, the World Bank, the European Bank for Reconstruction and Development (EBRD), the European Union (EU), and nongovernmental organizations (NGOs). The task force prepared a regional strategy for the elimination of gasoline lead by 2005 and set several intermediate targets. The strategy was broadly endorsed by the Fourth Environment for Europe Ministerial Conference, held in Århus, Denmark, in June 1998.

In connection with the UNECE task force and the Bank's support for the preparation of national environmental action programs (NEAPs), the World Bank was asked to assist the National Commitment Building Program to Phase Out Lead from Gasoline in Azerbaijan, Kazakhstan, and Uzbekistan. In the framework of this program, which was supported by the Danish Environmental Protection Agency (DEPA), preliminary studies were carried out to assess the level of lead pollution and to explore options for eliminating lead in gasoline in those three countries. The findings were discussed in May 1998 at a regional workshop in Almaty, Kazakhstan, which adopted a resolution stating that lead in gasoline should be eliminated by 2005 in Azerbaijan and Kazakhstan and by 2008 in Uzbekistan.

International experience shows that elimination of gasoline lead should be carried out within the broader context of fuel quality improvement and air quality management. A regional study conducted by the World Bank, Cleaner Transportation Fuels for Urban Air Quality Improvement in Central Asia and the Caucasus, accordingly takes this broader perspective. The study, which covered Armenia, Azerbaijan, Georgia, Kazakhstan, the Kyrgyz Republic, Tajikistan, Turkmenistan, and Uzbekistan, examined various factors affecting urban air quality and recommended cost-effective measures for improvements. This report encapsulates the results of the study.

Air quality monitoring, vehicle emissions inspection and maintenance, and fuel quality improvement form three closely interlinked facets of urban air quality management in the transport sector. Air quality monitoring data indicate which pollutants are exceeding national standards and health-based international guidelines. The choice of pollutants to be targeted for reduction will depend on their ambient concentrations, as well as their toxicity. The pollutants identified as posing a threat to public health in turn influence fuel parameters that would need to be tightened in cities where transport has been identified as a significant source of pollution. The costs incurred in improving fuel quality, however, will yield only limited benefits if vehicles using the fuels are not well maintained. This, then,

---

Box 1. International Experience with Phasing Out Gasoline Lead

Because of the extreme toxicity of lead, there is a worldwide move to ban leaded gasoline. More than three quarters of the gasoline sold today around the world is unleaded. The refining technology needed for switching from leaded to unleaded gasoline production is well known and commercially proven. By the end of 1999, about 35 industrial and developing countries had banned the use of lead in gasoline. Developing countries that have eliminated lead in gasoline include Bangladesh, Brazil, Guatemala, Hungary, India, the Slovak Republic, and Thailand. In Central Asia and the Caucasus region, Georgia banned the use of lead in gasoline effective January 2000.

Despite the significant progress made in getting rid of leaded gasoline, some misconceptions about unleaded gasoline persist. One is that only cars equipped with catalytic converters can use it. In reality, all gasoline-fueled cars can use unleaded gasoline. Another mistaken idea is that a large number of old vehicles running on leaded gasoline will suffer from valve-seat recession if they switch to unleaded gasoline. Laboratory tests have indeed demonstrated that in the absence of lead, which acts as a lubricant, valve-seat recession can occur if old vehicles with soft engine exhaust valve-seats are driven under severe conditions (that is, heavy loads and high speed). In practice, however, valve-seat recession has seldom been found to be a problem in most countries that have eliminated lead. In Latin America and the Caribbean, where lead phaseout has progressed rapidly in recent years, none of the countries have observed a marked increase in valve-seat problems.

In the early days of lead phaseout, the process lasted for decades in some countries—for example, in the United States. More recently, several countries, including the Slovak Republic and Thailand, have completed gasoline lead removal in four to five years. Countries that rely to a significant extent on imports can switch to unleaded gasoline even faster. Bangladesh and El Salvador, both with a refining industry, eliminated gasoline lead in less than a year. Such a swift transition to unleaded gasoline has a number of advantages. Little investment in the distribution infrastructure is needed because a dual distribution system does not have to be set up to segregate leaded and unleaded gasoline. Minimizing the transition period during which both leaded and unleaded gasolines are marketed also minimizes the chances of cross-contamination and misfueling of catalyst-equipped cars with leaded gasoline.

---

calls for having appropriate vehicle emission standards in place, identifying gross emitters, and requiring vehicles that fail emission tests to be repaired promptly.

In order to consider these three aspects of transport emissions control policy, the study assessed:

- The current status of air quality
- Current and future vehicle fleet characteristics and their fuel requirements
- The impact of different fuel specifications on vehicular emissions and air quality
- The implications of changing demand and fuel quality for the refining sector
- The technical feasibility and costs of various options for improving fuel quality
- Changes in petroleum sector policy, including pricing, fiscal measures, and liberalization of product trade to facilitate the introduction of cleaner fuels.

An examination of fuel specifications at the regional level is particularly timely for the following reasons:

- *Fuel specifications will have to be revised soon.* The old Soviet fuel standards can no longer meet the requirements of changing vehicle fleets and the imperative of protecting public health. All the NIS countries are reexamining the fuel and vehicular emissions standards they inherited from the former

Soviet Union. In March 1992 an interstate committee that included most countries in the region was established to deal with standardization, metrology, and certification. Heads of government signed an agreement accepting the standards of the former Soviet Union for a transitional period of undefined duration. Some countries have set up their own mechanisms to revise standards, but little progress seems to have been made to date.

- *Harmonization of fuel standards should be considered.* The rest of the world is moving toward harmonization of fuel and vehicle emissions standards. Harmonization makes for greater efficiency in vehicle manufacture and facilitates trade in refined products. Although the countries covered in this study are currently members of the interstate committee on standards and their fuel specifications are harmonized for the most part, there are now moves to set up country-specific standards, against the prevailing trend. The countries of Central Asia and the Caucasus could benefit from a consistent approach to fuel quality issues because of similarities in their urban air pollution problems and the potential for enhancing intraregional trade.

- *The refining industry needs guidance about future*

fuel standards so that it can plan its investments. The refining sector in the region is undergoing restructuring and privatization, and there are a number of proposals for refinery modernization schemes. Investors need clear signals from the government concerning future fuel specifications so they can optimize their investments, which are intended to have a life of about 20 years or longer. Without such guidelines, investment decisions may be less than optimal, or environmental objectives may suffer if there is resistance to changes in fuel specifications introduced after refinery modernization programs have started.

In recognition of the need for cross-sectoral cooperation, this regional study has drawn together government representatives from the environment, energy, and transport sectors, as well as from industry, academia, and NGOs. In addition to carrying out specific studies on urban air quality monitoring, vehicle emissions, and the downstream refining sector, the program has provided a forum for dialogue among policymakers, the private sector, multinational banks, aid agencies, and financiers on promoting a consistent approach to future formulation of standards and policies in the region. A more detailed description of the activities undertaken under the program is given in Annex A.

Air Quality Monitoring

Cost-effective reduction of pollution and human health damages requires an integrated approach to urban air quality management. An important step in developing an urban air quality management strategy is to be able to monitor and evaluate air quality. A good monitoring and modeling system is essential for policymaking suited to the primary objective of protecting human health. There are several key tasks for such a monitoring system.

Collecting data on ambient pollutant concentrations. The six most important pollutants to monitor regularly are what the WHO terms the "classical" pollutants:

- Lead
- $PM_{2.5}$/PM_{10} (particulate matter smaller than 2.5 and 10 microns in aerodynamic diameter, respectively)
- Carbon monoxide (CO)
- Sulfur dioxide (SO_2)
- Nitrogen dioxide (NO_2)
- Ozone.

It is good practice to monitor air quality at a variety of locations, including urban "hot spots" (areas affected by vehicular and industrial emissions), residential areas representative of population exposure, and rural areas (as an indication of background concentrations). The data will show which pollutants are exceeding national and international air quality standards and guidelines, such as the WHO's health-based air quality guidelines. These international standards and guidelines are for concentrations averaged over as long as a year, so continuous monitoring is important to make comparison possible.

Developing an emissions inventory. All sources of emissions, mobile and stationary, should be identified. Emissions are usually expressed in tons per year of pollutant. Measured in tonnage, CO typically leads all other pollutants in emissions levels, but it is important to bear in mind that its toxicity is orders of magnitude less than those of other pollutants. Some policymakers mistakenly add up tonnages of all the pollutants in the emissions inventory, note that CO emissions from vehicles are a sizable frac-

tion of the total, and conclude, for example, that "transport is responsible for 75 percent of air pollution." Such an approach does not take into account the toxicity, health impact, and dispersion of the various pollutants and may lead to erroneous conclusions about priorities.

Carrying out dispersion modeling. Dispersion modeling makes it possible to determine which emissions sources have the greatest effect on ambient pollution concentrations; this information is vital in developing an air quality management strategy. The main concern is human exposure. Emissions from tall stacks are dispersed much farther than those from low sources such as vehicles, and as a result the damage costs of emissions from vehicles far exceed those from tall stacks for the same tonnage.

CURRENT PROCEDURES

Historically, air quality monitoring was conducted throughout Central Asia and the Caucasus under the auspices of the Soviet Hydrometeorology Department. The status today varies from country to country, ranging from Uzbekistan, which still maintains an extensive network of operating monitoring stations, to Armenia and Georgia, where air quality monitoring ceased altogether in the late 1980s or early 1990s.

Where air quality monitoring procedures are in operation, they are identical across the region. The procedure, inherited from the Soviet era, consists of taking 20-minute readings, three times a day (typically, at 7:00 a.m., 1:00 p.m., and 7:00 p.m.), six days a week. Not all regulated pollutants are monitored. For example, ozone is monitored at only a handful of locations. Most monitoring stations appear to be measuring CO and total suspended particles (TSP, meaning particles of all sizes) regularly. Of particular importance for public health is that PM_{10} and $PM_{2.5}$ have not been monitored in the region, although they are far more damaging to public health than TSP (see Annex B). It is difficult to compare data collected on this schedule with the WHO guidelines, which typically call for maximal allowable levels for 1-, 8-, or 24-hour averages, as well as annual means.

As part of the World Bank study, air quality monitoring was conducted in Baku, Azerbaijan, and Tashkent, Uzbekistan, for about 10 days in each city in the summer of 1999. The purpose was to obtain data that could be directly compared with WHO guidelines and to compare the measurements with those collected by local hydrometeorology departments. NO_2, CO, ozone, and TSP were monitored using automatic, portable analyzers capable of collecting data continuously, and lead was collected using a personal monitor. Passive diffusion tube samplers were used to collect NO_2, SO_2, and ozone data over a wide area of each city (see Box 2).

RECOMMENDATIONS FOR AIR QUALITY MONITORING

The comparison between the World Bank sample monitoring in Baku and Tashkent and the hydrometeorology data showed that the departments underestimated CO concentrations on most occasions, overestimated TSP, and seriously underestimated lead concentrations. Even at today's very low levels of economic activity, the ambient concentrations of some pollutants already exceed the WHO guidelines. With the growth in the economies over the next 10 years, a business-as-usual scenario would lead to substantial exceedances of the guidelines. The limited data collected also suggest that pollutant concentrations are mismeasured in a number of cases. The following recommendations are suggested for improving the current monitoring system.

Equipment capability. The sampling strategy of monitoring for 20 minutes at a time three times a day is not considered effective for establishing mean or transient air quality indicators. It is especially unsuitable for areas where concentrations of pollutants change rapidly. Countries should strive to move away from the current system to

---

Box 2. Measuring Air Quality in Baku and Tashkent

To obtain data that could be used to assess current air quality monitoring procedures in Azerbaijan and Uzbekistan, portable high-resolution analyzers were located at hydrometerology observation posts, fairly close to major roads, in Baku and Tashkent. Under the weather conditions then prevailing, ambient pollutant concentrations would be expected to be low in Baku and high in Tashkent during the monitoring period.

Passive diffusion tubes, deployed at 20 locations in the two cities, recorded integrated average pollutant concentrations over the exposure period of about a week. In both cities, SO_2 levels were low. Ozone levels were much higher in Tashkent than in Baku.

Comparison with WHO guidelines. The data collected indicated that in Baku, over the course of 10 days, the WHO guidelines for NO_2 were exceeded twice. In Tashkent, over an 8-day period, WHO ozone guidelines were exceeded 15 times and NO_2 guidelines a total of 7 times. CO levels were well within the WHO guidelines in both cities. TSP levels were reasonable in Baku. (TSP could not be monitored in Tashkent because the TSP analyzer was damaged in transit.) Lead levels varied between 0.03 and 0.1 micrograms per cubic meter ($\mu g/m^3$) in Baku and 0.15 $\mu g/m^3$ in Tashkent.

Comparison with current monitoring methods. Comparison of the data obtained using automatic, continuous analyzers with that gathered by the country hydrometerology departments showed a range from very good agreement to differences of severalfold. The NO_2 values obtained using continuous analyzers differed from those obtained by the hydrometerology departments by up to a factor of three; those for CO differed by up to a factor of five and those for TSP and lead by an order of magnitude. It appears that the level of airborne lead has been historically severely underestimated in this region. Since monitoring procedures are uniform throughout Central Asia and the Caucasus, these findings indicate similar underestimations over a wider area. Some selected data are presented in Annex C.

Most problematic, perhaps, are the measurements of particulate matter. Comparisons of CO and TSP concentrations were complicated by the fact that the results reported by the hydrometerology departments are at the detection limits for the techniques used. For example, in the case of TSP, the resolution of the balance used is poor, so that the reported values are at the detection limits for the balance. In addition, there appears to be inadequate filter conditioning, and the filters used are not suitable for collecting particles at the prevailing flow rate. Furthermore, the equipment currently in use cannot measure PM_{10} or $PM_{2.5}$, the particles of greatest concern for public health.

continuous monitoring, which will make it possible to assess compliance with the WHO air quality guidelines. In the current setup, lead can be monitored only by collecting particulate filters for a month. It is important to be able to obtain time-resolved lead data rather than monthly averages.

Selection of pollutants to be monitored. Regular measurement of PM_{10} or $PM_{2.5}$ would be much more informative than measurement of TSP. It is the public health impact of air pollution that is of most interest, and coarse particles, which are part of the TSP measurement, have been found not to have measurable health effects. Ozone has been measured at only a single observation post in Tashkent, where the study showed that WHO guidelines were being exceeded. Given the results from the continuous monitoring and passive sampling studies, ozone should be measured at additional observation posts in cities with indications of high ambient ozone concentrations. In the initial stage of monitoring system modification, governments may consider reducing the number of pollutants that are regularly measured in order to offset the incremental cost of modernizing the equipment.

Site selection. Because most monitoring stations are currently located in residential areas, little information can be obtained about pollutant concentrations at "hot spots" or about background concentrations. In the long run, it would be useful to deploy a number of sites at curbside, industrial, and rural locations to build a fuller picture of air quality.

The cost implications of these recommendations need not be high. Even in the most comprehensive program, a continuous air quality monitoring station for measuring the six criteria pollutants can be set up for less than US$300,000. In the early stage of system modification, it may be wise to limit the number of pollutants and target, for example, only fine particles and lead. Portable continuous monitors for measuring PM_{10} are available for US$3,000. The cost of setting up and operating one continuous station monitoring the six classical pollutants, supported by several satellite sites monitoring only PM_{10} and lead, for three years may be on the order of US$1.5 million to US$2.5 million.

The Vehicle Fleet and Vehicle Technology

Two major changes expected to take place in Central Asia and the Caucasus in the coming decade are the replacement of heavy-duty gasoline vehicles by diesel vehicles and the phase-out of low-octane passenger cars. Both changes will have a significant effect on the amount and type of fuel used and on vehicular emissions. The growing use of diesel vehicles will mean increased emissions of fine particulate matter and oxides of nitrogen (NO_x). Decreased use of low-compression-engine gasoline vehicles will raise fuel economy, resulting, in principle, in the lower emissions of CO, hydrocarbons, and NO_x typical of gasoline vehicles and in lower greenhouse gas emissions for the same distance traveled.

CHARACTERISTICS OF THE VEHICLE FLEET

The numbers of vehicles in the region range from about 300,000 in smaller countries to 1.5 million in Kazakhstan. In most countries, heavy-duty vehicles constitute about one fourth of the total vehicle fleet. Although detailed data on the age distribution of vehicles are not readily available, the average age appears to be high, with a noticeable number of vehicles 20 years old or older. The total number of vehicles in the fleets has not increased much in recent years and has even declined somewhat as a consequence of the general state of the economy of these countries in the 1990s. The official vehicle statistics from vehicle registration are given in Table 1.

One of the most important factors in achieving reduced emissions from privately owned vehicles is the availability of service and repair facilities with good diagnostic equipment and qualified technicians. At present, the availability of such facilities in Central Asia and the Caucasus is limited. Vehicle maintenance service has been difficult to obtain even in major cities and is virtually nonexistent in the countryside. In most instances in the

TABLE 1. VEHICLE STATISTICS, CENTRAL ASIA AND THE CAUCASUS, 1998

| Country | Highways (kilometers) | | | Total vehicles | Total light duty | Total heavy duty |
	Paved	Unpaved	Total			
Armenia[a]	8,560	0	8,580			
Azerbaijan	54,188	3,582	57,770	365,782	272,092	93,690
Georgia	19,354	1,346	20,700	406,733	340,407	66,326
Kazakhstan[b]	104,200	36,800	141,000	1,496,969	1,098,548	398,421
Kyrgyz Republic[a]	16,854	1,646	18,500			
Tajikistan	11,330	2,370	13,700	329,996	120,819	209,177
Turkmenistan	19,488	4,512	24,000	314,990	252,082	62,908
Uzbekistan	71,237	10,363	81,600	1,139,849	889,286	250,563

a. Vehicle fleet data from Armenia and the Kyrgyz Republic were not available at the time of publication.
b. Kazakhstan has 67,630 vehicles categorized as "other."

past, a good set of tools came with the vehicles, and maintenance or repair was performed by the owner. Vehicle designs were kept simple and consistent.

Most of the vehicles currently used in Central Asia and the Caucasus were manufactured within the former Soviet Union during the 1980s. Heavy-duty vehicles use relatively small gasoline engines, in contrast to the dominance of diesel engines in other parts of the world. These are low-compression engines that operate on low-octane gasoline, typically 76 motor octane number (MON)[2]. They have poorer fuel economy, lower energy efficiency, and higher emissions levels of carbon dioxide (a greenhouse gas) and other pollutants than high-compression engines for the same distance traveled. The gasoline-powered vehicles utilize conventional carburetor technology and do not incorporate emissions control systems such as catalytic converters. The typical domestic vehicle technology is similar to what was common in North America and elsewhere until the mid-1980s.

Leaded Gasoline: Some Issues

The average octane requirement has been increasing in the region in recent years as a result of the introduction of modern engines, but the average octane of the gasoline actually available on the market remains low because of the limitations of domestic refineries and the costs incurred in increasing the octane. In some countries this has given incentives for the illegal addition of lead to gasoline at the retail level, with serious adverse effects on public health.

Catalytic converters are by far the most effective means of reducing emissions of CO, hydrocarbons, and NO_x from gasoline vehicles. However, lead permanently poisons catalysts, and currently the gasoline distribution system in the region does not segregate leaded and unleaded gasoline. The introduction of catalytic converters should await the complete phaseout of lead in gasoline so that catalyst-equipped vehicles will not be misfueled with leaded gasoline.

Because of the widespread use of hardened valve-seats in vehicles throughout the NIS, the elimination of lead in gasoline should not cause valve-seat recession. Indeed, in Azerbaijan, where gasoline has been entirely unleaded since 1997, there are no reports of a marked increase in valve-seat recession problems. Elimination of lead yields certain benefits for consumers: longer engine and exhaust valve life, much longer exhaust system life, and less frequent oil changes and spark plug replacement. It is important to educate the general public about these benefits and to clear up the misperceptions mentioned in Box 1.

Future Fuel Consumption and Octane Requirements

The gasoline-powered vehicles that currently make up a large part of the heavy-duty fleet were mostly designed for 76 MON gasoline. These vehicles will eventually be

---

Box 3. Calculating Trends in Fuel Consumption

To estimate the octane requirements of the current vehicle fleet and future fuel consumption, a computer model was developed that incorporates vehicle types, vehicle models, manufacturer-specified fuel octane requirements, fuel economy (kilometers traveled per liter), and estimates of annual vehicle kilometers traveled (VKT). The information came from local contacts and from the literature. The fuel requirements for any given year are calculated by using estimates of annual VKT by vehicle class—for example, 40,000 kilometers (km) for trucks, 50,000 km for buses, 15,000 km for cars, and 10,000 km for special vehicles. The same annual VKT figures were used for all eight countries.

The calculations were carried out for each vehicle type, and results were obtained for three fuel categories: low-octane gasoline [76 MON, 80 RON (Research Octane Number)], high-octane gasoline (91 RON and higher), and diesel. For projections to 2005 and 2010, the model assumes a 5 percent per year reduction in low-octane gasoline usage by heavy-duty vehicles and by light-duty vehicles that use low-octane gasoline. In sensitivity analysis, an annual replacement rate of 7.5 percent instead of 5 percent was tested. Diesel and high-octane gasoline usage by the rest of the fleet was adjusted by correlating growth in total VKT with the forecast growth of GDP in each country. The growth rate of total VKT was made directly proportional to that of GDP in the base case. Sensitivity analysis included changing the proportionality constant to 1.25.

---

[2] MON is a measure of resistance to self-ignition (knocking) of a gasoline when vehicles are operated under conditions that correlate with road performance during highway driving conditions.

replaced by diesel engines. According to information from the transport ministries in the region, diesel fuel use in the heavy-duty sector is increasing as a result of the lower cost of diesel and the greater durability of diesel engines. The rate of diesel use by the heavy-duty vehicle fleet has a significant impact on the fleet octane mix. The study included estimations of future fuel consumption (Box 3).

Figures 1 and 2 show the cumulative percentage growth rates for gross domestic product (GDP), gasoline consumption, and on-road diesel consumption between 1998 and 2005 and between 1998 and 2010, using the base-case calculations. As a result of the switch from gasoline to diesel, the consumption of gasoline is almost static in most of the countries, and its growth is negative in Tajikistan and Uzbekistan. Diesel use grows very rapidly—in the case of Turkmenistan and Uzbekistan, as much as four times more than GDP between 1998 and 2010. This would significantly increase the emissions of fine particulate matter and NO_x in the coming years. Figure 3 shows the estimated percentage of high-octane gasoline in total consumption in 1998, computed on the basis of vehicle fleet inventory and vehicle manufacturers' octane recommendations, and forecasts for 2005 and 2010 based on manufacturers' recommendations.

It was not possible to obtain accurate information on the current vehicle fleet inventory. Vehicles that have long

been retired are still registered; estimates of the share of vehicles that are registered but not operating are as high as 40 percent in some countries. It is known that aftermarket remachining of gasoline engines to a lower compression ratio occurs, but no statistics are available for the number of cars affected. No systematic data have been compiled on annual VKT or on actual fuel economy. All these limitations on data availability introduce a large margin of uncertainty into the calculations.

Heavy-duty vehicles are significant consumers of gasoline because of their low fuel economy and their high annual VKT in comparison with other vehicle categories. The switch from gasoline to diesel by these vehicles means that gasoline demand in the region is not expected to grow much in the foreseeable future. In all cases, including those run under sensitivity analysis, growth of demand for gasoline is lower than GDP growth. There is a caveat, however: if the number of heavy-duty gasoline-powered vehicles in the model is overestimated relative to other categories of vehicle, the results presented above will exaggerate the decrease in gasoline demand in the future from this segment of the vehicle population.

The gap between the actual octane used and the octane requirements specified by vehicle manufacturers varies markedly from country to country. On the basis of the limited information available in Azerbaijan and Georgia, many vehicles designed for low octane may be using high-octane

FIGURE 1. PROJECTED CUMULATIVE GROWTH OF GDP, GASOLINE CONSUMPTION, AND DIESEL CONSUMPTION, 1998–2005

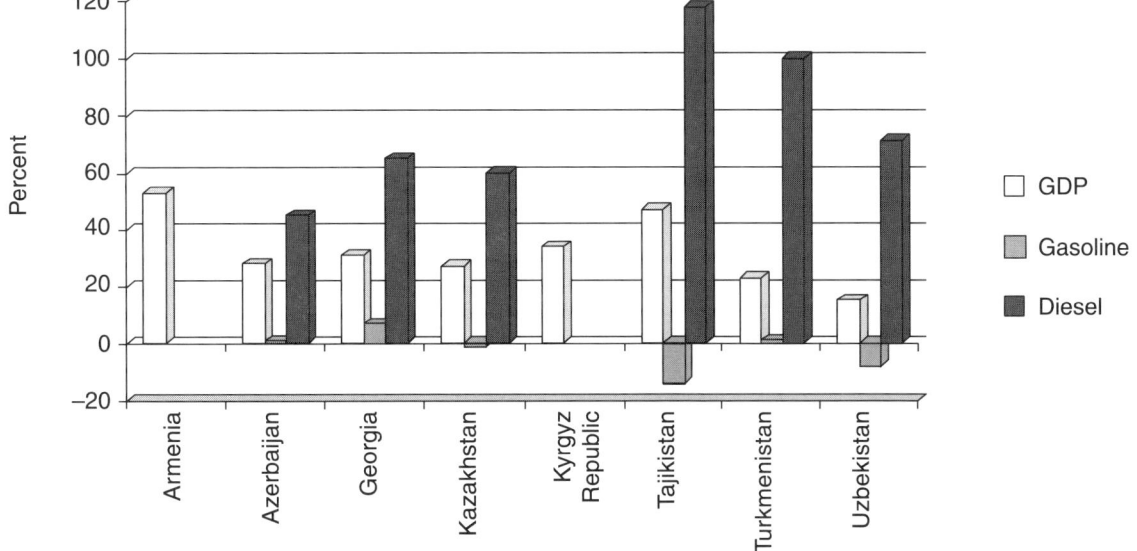

Note: GDP growth for Azerbaijan does not assume timely completion of export oil pipelines. Because vehicle fleet data for Armenia and the Kyrgyz Republic were not available, no calculations could be carried out.

Sources: For GDP, the World Bank; for gasoline and diesel consumption, Environment Canada.

FIGURE 2. PROJECTED CUMULATIVE GROWTH OF GDP, GASOLINE CONSUMPTION, AND DIESEL CONSUMPTION, 1998–2010

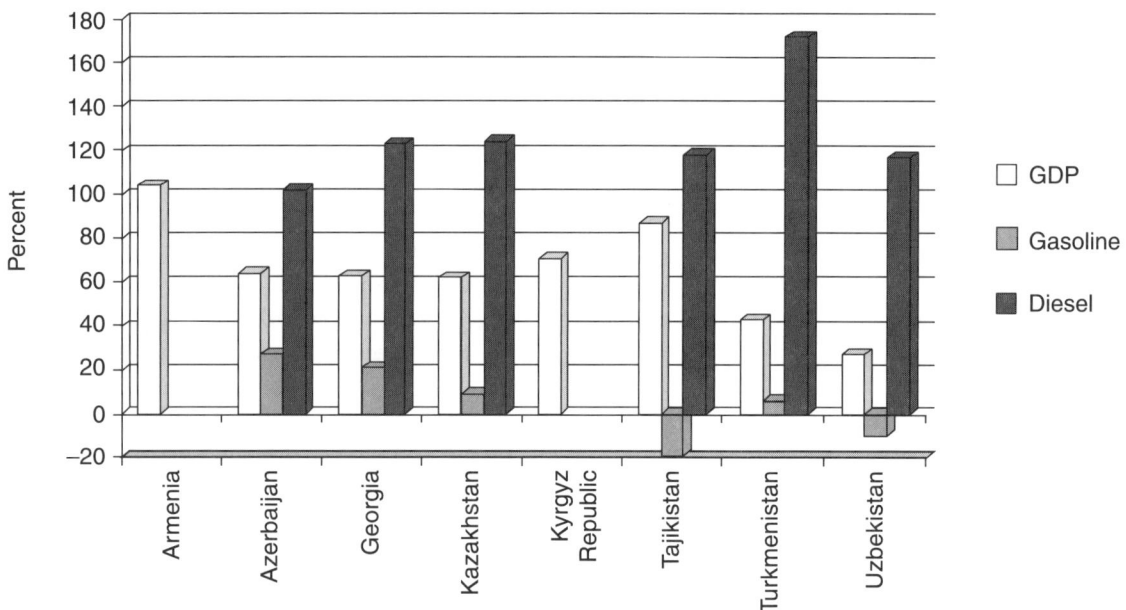

Note: GDP growth for Azerbaijan does not assume timely completion of export oil pipelines. Because vehicle fleet data for Armenia and the Kyrgyz Republic were not available, no calculations could be carried out.

Sources: For GDP, World Bank; for gasoline and diesel consumption, Environment Canada.

FIGURE 3. SHARE OF HIGH-OCTANE GASOLINE IN TOTAL GASOLINE CONSUMPTION, 1998, 2005, AND 2010

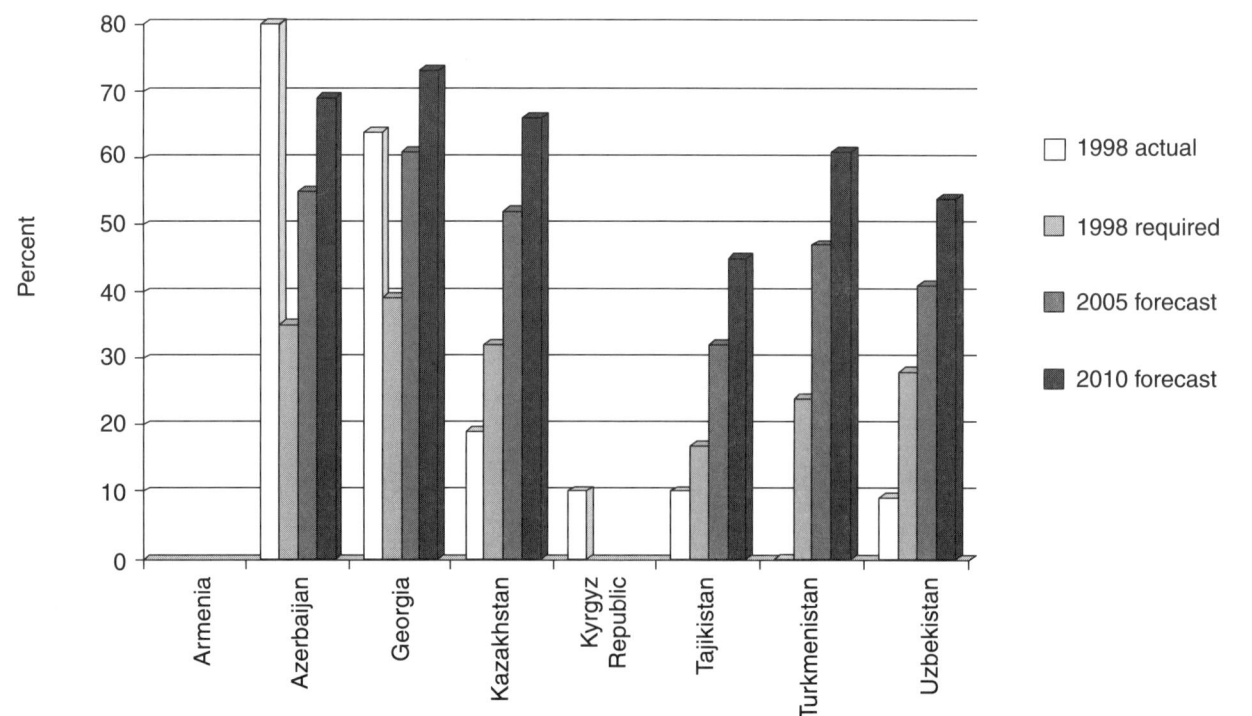

Note: The only data available in Azerbaijan and Turkmenistan were refinery production figures, not consumption figures as a function of octane. No octane split figures were available in Armenia. The 1998 actuals for Georgia, the Kyrgyz Republic, and Tajikistan are estimates.

gasoline in these countries. There is no harm in running low-compression-engine vehicles on high-octane gasoline, although there is no benefit to the driver, either, and higher-octane gasoline is more costly. The high percentages of high-octane gasoline consumed may be overestimates, since the Azerbaijan figure is based on refinery production only, and it is known that gasoline is smuggled into areas of Azerbaijan near the border. Estimates for the amount of gasoline smuggled into Georgia run as high as 50 percent, depending on the international price of gasoline.

The rest of the region shows a significant octane shortfall: the amount of high-octane gasoline consumed is lower, and in some cases considerably lower, than that computed on the basis of vehicle fleet inventory and manufacturers' recommendations. This gap will widen in the future as vehicles using low-octane gasoline are replaced and retired and as elimination of lead requires that average octane be raised. This has implications for policy toward lead emissions because although the consumption of gasoline may not increase, the amount of lead that might be used to achieve the target octane level would be substantially higher. The countries in the region face a significant challenge, particularly those with a sizable refining sector. The rising octane requirement of the vehicle fleet is an additional burden on the refineries, which are already octane short.

VEHICLE EMISSIONS MONITORING

Data from around the world indicate that vehicle technology and vehicle performance generally have a much greater impact on emissions levels than does fuel quality (except for lead and sulfur, where the amount emitted is directly proportional to the amount in the fuel). This is particularly true of particulate, hydrocarbon, CO, and NO_x emissions. In this context, proper vehicle maintenance plays a crucial role in minimizing vehicular emissions. One way of enforcing regular vehicle maintenance is to carry out periodic emissions inspection and require that in-use vehicles meet the emissions standards.

Exhaust emissions measurements of CO and hydrocarbons for gasoline vehicles, and of opacity for diesel vehicles, are part of the required periodic technical inspections for road vehicles in the NIS. For gasoline-fueled engines, the CO emissions standards (3 percent at idle) can effectively identify gross maladjustment of the air-fuel ratio. Use of these standards can help reduce excess emissions if regularly implemented for older vehicles with carburetors and no emissions control devices. Most of the vehicle fleet in Central Asia and the Caucasus conforms to this descrip-

tion. Although the hydrocarbon emissions limits are disproportionately high, it is the CO measurement equipment that is more commonly available, and hence the CO values represent the de facto emissions limit. For diesel vehicles, exhaust opacity measurements during snap acceleration are taken. This is the same approach used as the primary method of emissions control in North America. Thus, it would appear that a satisfactory inspection and maintenance (I/M) system is, in principle, in place, provided that implementation and enforcement of these measurements are properly carried out.

Although regulations that call for measurement as frequently as every three months are on the books, observations suggest that an effective I/M program based on these regulations is not necessarily operating.[3] Because neither adequate equipment nor the trained personnel to operate it is generally available, many fewer tests are performed than would be expected if the regulations were enforced rigorously. Before the breakup of the Soviet Union, the Soviet Standards Institute provided assistance with the annual calibration of emissions measurement equipment and the training of personnel. Since then, links with the institute have become weaker, and new institutions have not been established.

The lack of detailed records from I/M programs is a strong indication that these programs fall short of the written regulations. Records on the numbers of vehicles tested and of vehicles that passed, if kept at all, appear to be available only for the entire vehicle fleet, without breakdown by type or age. Keeping relatively simple statistics of pass/fail rates for different ages and models of vehicles would be a first step toward determining whether these I/M programs have any impact on ambient air quality. The information gathered from regularly scheduled tests can be used to target the right vehicles (gross emitters) during roadside checks so that the limited number of checks can be more effective.

The consequence of failing an emissions test in a periodic or roadside check is a monetary fine that is collected on the spot without corrective action necessarily being taken. Although the fines are typically not trivial, they may be less than the cost of having the vehicle serviced or than an alternative informal fine. The payment of the fine, rather than corrective action that might reduce the excess emissions, is often regarded as the central issue.

VEHICLES POWERED WITH ALTERNATIVE FUELS

Vehicles that use alternative fuels are a very small portion of the fleet and are typically retrofits of in-use gasoline

---

[3] The three-month interval is unrealistic and is, moreover, unnecessary for passenger cars.

engines. Before the breakup of the Soviet Union, no dedicated alternative-fuel-powered vehicles were being manufactured except for specialized light-duty trucks used by city services. Among the countries in Central Asia and the Caucasus, Uzbekistan, which has significant natural gas resources, is particularly interested in use of compressed natural gas (CNG) in vehicles. Uzbekistan has plans to expand the use of CNG in the vehicle fleet, including light-duty vehicles, and to introduce liquefied natural gas (LNG) usage in special niches such as railroad locomotives and heavy-duty mining vehicles.

The share of CNG in meeting Uzbekistan's transport energy consumption is 5 percent, and there are 33 CNG fueling stations. Although the existing stations are currently operating well below their estimated capacities, another 32 fueling stations are planned or are under construction to serve an expanded CNG fleet.

CNG vehicles in Uzbekistan are nearly all heavy-duty vehicles and buses, representing 4.3 and 3.6 percent of the respective fleets in 1995 (USTDA 1997). Most of these vehicles have been converted to CNG from gasoline; vehicles that have been converted from diesel use a mixture of diesel and CNG. The fuel kits for conversion to CNG are imported from Russia. The fuel tanks, 8 to 10 per vehicle, are made of heavy steel, adding 600–1,000 kilograms (kg) to the weight of the vehicle and giving a typical driving range of 250 km.

CNG conversions are at present dormant; in 1997–98 only 168 vehicles were converted. For comparison, 1,430 vehicles were converted in 1995, and there is a countrywide conversion shop capacity of nearly 12,000 vehicles per year (Pyadichev 1999). The current state of conversion activity probably reflects the limitations of the existing technology (steel tanks and open-loop fuel conversion kits from Russia). The implementation of a large-scale conversion program will require investments in Uzbekistan for producing lightweight fuel tanks made of composite materials and electronically controlled conversion kits that can operate with closed-loop electronic fuel and air management systems. Future facilities for producing pressure reduction and mixing devices, composite fuel tanks, and electronic components have been approved by government authorities. It is estimated that it will be two years before production of composite fuel tanks begins. Discussions are presently being held on financing a composite fuel tank factory.

RECOMMENDATIONS FOR MONITORING AND ABATEMENT OF VEHICULAR EMISSIONS

Equipment upgrade. It is not clear that adding more I/M equipment under the existing programs will translate into a more effective I/M program. Reduction of emissions from the vehicle fleet depends on the general state of vehicle maintenance infrastructure and the feasibility of new opportunities for cost-effective solutions to I/M tasks. Thus, equipment upgrade, while a high priority, should be accompanied by the expansion of service and repair facilities.

Service and repair facilities. Among the most important factors in achieving reduced emissions from the privately owned vehicle fleet is the availability of service and repair facilities with good diagnostic equipment and qualified technicians. Centralized I/M programs with "test only" facilities are generally viewed as more effective than decentralized I/M programs with "test-and-repair" facilities because of the potential for conflict of interest in the latter case. However, if adequately equipped and staffed service and repair facilities are not readily available, a rigorous "test only" program may well prove to be little more than a mechanism for collecting fines. Vehicle maintenance will be increasingly crucial as consumption of diesel fuel in the transport sector in the region rises rapidly in coming years. Diesel particulate emissions have been found to be particularly injurious to human health (see Annex B), and poorly maintained diesel vehicles are responsible for high levels of particulate emissions. In order to create demand for upgrading and expanding service and repair facilities, governments should start enforcing emissions standards more rigorously by requiring that vehicles that fail be repaired and retested. Requiring motorists to pay for regular repair and maintenance upholds the "polluter pays" principle.

Vehicle registration. Reliable vehicle registration records are a prerequisite for mounting an effective I/M program. An up-to-date database, computerized to enable searches, with retired vehicles eliminated from the records, will help track gross polluters for targeting, thus making the most efficient use of the limited resources available for emissions testing. Establishment of such a database is a relatively straightforward task that can be accomplished with modest marginal resources and would have a positive effect on all record-keeping activities.

Remote sensing for light-duty gasoline vehicles. Given the present resource limitations, it may be worthwhile to consider alternatives to the "test every vehicle" principle. One option is to screen a large number of vehicles to identify gross emitters for more intensive testing. Light-duty gasoline vehicles can be screened by means of remote-sensing technology, which utilizes infrared spectroscopy to measure concentrations of hydrocarbons, CO, and NO_x in the exhaust plume of a vehicle while it is being driven on the road. The speed and acceleration of the vehicle and an image of the license plate can be recorded simultaneously, making it possible to identify vehicles and determine the conditions under which the measurement was taken. Commercially available remote-sensing devices can measure more than 4,000 cars per hour on a continuous basis, providing a powerful tool for characterizing emissions from the on-road vehicle fleet. Remote sensing has been successfully used as a component of I/M programs in North America, either to identify high-emitting vehicles and call them back for repair or to identify clean vehicles and exempt them from regularly scheduled periodic measurement at I/M stations. Although remote sensing has been widely used only in North America, it offers a cost-effective solution to vehicle fleet screening in developing countries. During 1997–99 Taiwan (China) implemented a remote-sensing program using 10 stations. That experience might be interesting to evaluate should a pilot project be considered in Central Asia and the Caucasus. Other metropolitan areas such as Beijing, Bangkok, and Seoul are planning similar remote-sensing programs.

Estimates of US$90,000 to US$140,000 have been made for stationary and mobile remote-sensing stations, including training of personnel. More precise costing and arrangements for such a pilot project would need to be negotiated with a commercial vendor. Although the costs are significant, the relevant comparison is with the cost of setting up a large number of traditional I/M stations that can handle comparable numbers of vehicles.

Utilizing remote sensing as an I/M or screening tool requires an efficient database of vehicle identification records so that license plates can be matched with other details of the vehicle. As noted above, such a database is necessary, in any case, for an I/M program that is effective in reducing air pollution.

Downstream Petroleum Sector and Fuel Quality

The countries of Central Asia and the Caucasus may be broadly divided into two categories: those with a significant refining capacity (Azerbaijan, Kazakhstan, Turkmenistan, and Uzbekistan), and those without one. Importing countries—the latter group—generally find it easier to improve fuel quality, as they do not have to make substantial capital investments to modernize refineries. It is not surprising that Georgia, the only country in the region to have officially banned the use of lead in gasoline, imports almost all of the gasoline it uses. Among the countries with large refin-

ing capacity, Azerbaijan, Turkmenistan, and Uzbekistan are self-sufficient in the supply of gasoline and diesel. Kazakhstan has sufficient refining capacity to be self-sufficient in gasoline, but the refineries have been running at a relatively low rate of utilization, and as a result, the country has been importing a significant amount of gasoline.

The refineries in the region vary in configuration from small topping refineries (the simplest configuration possible) to complex refineries with cracking units. Among the complex refineries are the Pavlodar refinery in Kazakhstan, the Nova Baku refinery in Azerbaijan, and the Turkmenbashi refinery in Turkmenistan. The locations of some of these refineries are not optimal. Most of them were built during the Soviet era, when the supply of crude and product trade were controlled and guaranteed by the government. Since the breakup of the Soviet Union, some refineries have been cut off from sources of crude or have found it difficult to sell to markets that are now in other countries.

A sizable number of large refineries are currently running at a level considerably below capacity. In some cases the reason is that the refineries, which in the Soviet era had guaranteed access to crude, are now having to import and pay for it. The Pavlodar refinery in Kazakhstan has had a particularly difficult time securing crude and has had a very low rate of utilization in recent years. The Chardzou refinery in Turkmenistan, which was designed to use crude from Uzbekistan, has also had trouble securing crude in recent years and as a result has not operated much of the time. Another reason for low utilization is the steep decline

in the demand for refined products since the early 1990s as a result of the contraction of the economy. The Nova Baku refinery in Azerbaijan, for example, has faced a marked fall in gasoline demand, and recently the fluidized catalytic cracking (FCC) unit has seldom been in operation. The refinery staffs report that gasoline demand will have to nearly double before the FCC unit can start running again on a regular basis.

The refineries in Azerbaijan, Turkmenistan, and Uzbekistan are under government control, although the government of Uzbekistan is in the process of privatizing the Fergana refinery. In large crude-producing countries such as Azerbaijan and Kazakhstan, interest for the private or government sector lies primarily in upstream operations, making it difficult to attract capital to refineries for modernization. The most significant modernization program under way today is at the Turkmenbashi refinery in Turkmenistan, which is completing Phase 1 of its program at a cost of about US$1 billion. Several Japanese firms are considering a US$0.5 billion upgrading for the Atyru refinery in Kazakhstan, and the Shymkent refinery, also in Kazakhstan, is looking for financing to support its upgrade scheme.

THE QUALITY OF TRANSPORT FUELS

Fuel quality and vehicle technology interact closely and influence each other's impacts on vehicle emissions, fuel consumption, and vehicle durability. A few pollutants—lead, heavy metals, and SO_2—can be controlled through fuel quality alone. By eliminating lead and other heavy metals in gasoline and reducing sulfur in gasoline and diesel, the ambient concentrations of these pollutants can be significantly lowered. The emissions levels of all other pollutants depend on both fuel quality and vehicle technology. (See Annex D for a more detailed description of the impact of fuel quality on vehicle emissions.)

The fuel quality parameters that should be considered in dealing with air quality management in Central Asia and the Caucasus include the amounts of lead, sulfur, aromatics, benzene, olefins, and oxygenates in the fuel; gasoline volatility; and distillation control. The type of crude processed, refinery configuration, the severity of refinery process unit operation, and product slate all affect these fuel parameters.

Current Specifications

Fuel specifications in the region are evolving, but for the most part the standards set during the Soviet era are in force.

- *Lead.* The use of lead in gasoline has been banned only in Georgia to date, although leaded gasoline has not been produced in Azerbaijan since 1997. In all the countries except Georgia, up to 0.17 grams of lead per liter of gasoline (g/l) is allowed in 76 MON gasoline and 0.37 g/l in 93 research octane number (RON)[4]. The exception is in cities with a population of over 1 million inhabitants, where the use of lead in gasoline has been banned since the 1980s, although compliance has not been strictly enforced.

- *Octane.* The lowest octane grades available are 72 MON and 76 MON. Their corresponding research octane numbers are in the vicinity of 80. Although vehicles that were manufactured to run on low-octane gasoline are still found in the region, all modern-engine vehicles require a minimum of 91–92 RON. As vehicle fleet renewal proceeds, these low-octane grades should be phased out.

- *Benzene and aromatics.* There are currently no limits on the benzene or aromatics content of gasoline in the region. Elsewhere, awareness of the cancer-causing potential of benzene has led to its limitation. The Russian Federation limited benzene in unleaded gasoline to 5 volume percent (vol%) beginning in January 1999, and the EU lowered the maximum permissible level of benzene in gasoline from 5 vol% to 1 vol% effective January 2000. For the countries of Central Asia and the Caucasus, the question is how to phase in such increasingly stringent fuel specifications.

- *Ozone.* Limits on Reid vapor pressure (RVP) control gasoline volatility and hence evaporative emissions. The summertime (ozone-season) limit on RVP in the region is 66.7 kilopascals (kPa). Lowering the RVP is one of the most cost-effective ways of controlling ozone, as it reduces photochemically reactive hydrocarbon emissions. Cities in which ground-level ozone is a problem may wish to consider this step.

- *Sulfur.* Sulfur in gasoline is limited to 0.1 weight percent (wt%). Since sulfur is a temporary poison for catalysts, countries should consider tightening the limit as the numbers of gasoline vehicles equipped with catalytic converters increase in the region.

- *Diesel fuels.* Key diesel specifications include a minimum cetane requirement of 45, maximum sulfur of 0.5 wt% for Type II and 0.2 wt% for Type I, and summertime T96 (the temperature at which 96 percent of diesel evaporates) of 360° Celsius. The cetane, T96, and Type I sulfur standards are reasonable. If Type II diesel is used in transport, sulfur reduction should be considered.

As part of the study, samples of fuel were tested against the specifications in force (Box 4).

Fuel Quality Monitoring

Fuel quality monitoring efforts in the region are limited. The most serious problem appears to be the adulteration of transport fuels downstream from refineries. Common practices include the adulteration of gasoline with (lower-priced) kerosene and the addition of lead and other heavy metal additives to gasoline to increase octane. Informal addition of lead to gasoline in an uncontrolled environment poses a serious health risk to those involved in such an activity. One of the gasoline samples taken in this program could not be analyzed for lead because of the presence of excess iron, probably in the form of ferrocene.

A concerted effort to monitor gasoline lead more rigorously is being made in Almaty, which established an interdepartmental commission on gasoline quality in July 1998. Monitoring of gasoline lead between 1997 and the beginning of 1998 indicated that as much as one third of the gasoline sold was leaded, although in Almaty, with its population of more than a million, only unleaded gasoline is supposed to be sold. The commission reports that it has been successful in tackling centralized supply of lead through railroad terminals, which receive bulk deliveries and are more easily checked, and is now addressing com-

---

[4] RON is a measure of resistance to self-ignition (knocking) of a gasoline when vehicles are operated at low speed under city driving conditions.

Box 4. Results of Fuel Sample Testing

In the framework of the regional study, gasoline and diesel samples were taken in a number of cities throughout the region, in every country except Turkmenistan. The most extensive sampling was conducted in Azerbaijan, where 15 gasoline samples and 7 diesel samples were tested. The results indicated that the quality of gasoline and diesel varied considerably across the region.

The specifications for gasoline octane were violated frequently. In one country, five out of six 76 MON gasoline samples did not meet the minimal MON requirement. In another country, two samples of 76 MON gasoline tested had only 75 RON/70 MON. One sample of A93 had 83 RON/78 MON. There were indications that some gasoline samples were mixed with kerosene.

The level of lead in gasoline did not exceed the maximal permissable levels, with the exception of one sample of 76 MON that contained 0.3 grams per liter (g/l) and another in a city with a population of over 1 million where the legal limit of 0.013 g/l was exceeded. In some cases, large quantities of iron additives were found in gasoline, suggesting aftermarket addition downstream of refineries.

High-octane gasoline contained high levels of aromatics—as high as 59 wt% aromatics and 5.3 wt% benzene. This is indicative of a refining configuration that relies primarily on reformate as the principal source of octane. Lowering the level of benzene and aromatics should be a priority for the protection of public health.

The level of sulfur in gasoline exceeded the legal limit in several cases. The highest level found was 0.14 wt% (the legal limit is 0.1 wt%).

Diesel samples met all the specifications, for the most part, except for sulfur. The highest amount of sulfur in diesel found was over 0.9 wt%, recorded in two countries. In most countries diesel sulfur was at or below 0.2 wt%. There were signs of cross-contamination of diesel with crude oil.

---

pliance by trucks and the sale of leaded gasoline at the retail level.

The results of limited fuel monitoring conducted under this program, taken together with observations by local specialists, indicate a clear need to significantly strengthen fuel quality monitoring. The steps involved in setting up an effective system are to examine the regulatory framework, including handling of noncompliance; identify key parameters to monitor; determine which body should do the monitoring; provide for quality assurance and quality control in the laboratories that test fuel samples; and decide on the statistical procedure for sampling—where, how often, and when.

CLEANER FUELS AND REFINERY CONFIGURATIONS: OPTIONS FOR THE REGION

As fuel specifications become tighter, it is increasingly difficult for small, simple refineries (topping or hydroskimming types) to meet new fuel specifications because of the economies of scale involved in constructing various processing units. Some "mini"-refineries in the region—all small topping refineries—would not be able to supply unleaded gasoline at a competitive price. The only option for these refineries in the face of lead elimination would be to blend straight-run naphtha with high-octane gasoline or blending components to make 76 MON gasoline. The role of such refineries should be reexamined in the future. (Box

5 discusses some of the cost implications of efforts to improve quality.)

The principal challenge facing the larger refineries in Azerbaijan, Kazakhstan, Turkmenistan, and Uzbekistan is how to phase out lead and increase average octane without increasing benzene and total aromatics to unacceptably high levels. Limiting aromatics rules out the possibility of relying only on reformate to compensate for the octane shortfall. Other sources of octane, such as isomerate and oxygenates, will be needed. Installing an isomerization unit is relatively inexpensive. Oxygenates will have to be purchased from ether or alcohol suppliers.

Three refineries have FCC units: Nova Baku in Azerbaijan, Pavlodar in Kazakhstan, and Turkmenbashi in Turkmenistan. FCC naphtha is an alternative source of octane. However, FCC units often do not run because of the depressed demand for domestically produced gasoline or because of difficulties in securing crude.

Turkmenbashi is expected to be in a position to phase lead out completely and to limit benzene and total aromatics to a reasonable level once the current reconfiguration project is completed. All of the remaining refineries need to take steps to lower the content of benzene and total aromatics. Options include installing a naphtha splitter upstream of the reformer to fractionate benzene precursors, installing an isomerization unit, and purchasing ethanol or ethers such as methyl tertiary butyl ether (MTBE). For refineries with FCC units, alkylation is another option, although the necessary unit is expensive to build.

Depending on the type of crude processed, reduction of sulfur in diesel fuel may be necessary. This requires hydrodesulfurization. The Fergana refinery in Uzbekistan, which processes high-sulfur crude, has installed a diesel hydrotreater and should be in a position to meet the diesel sulfur specification after the hydrotreater comes on stream.

While the final cost to consumers of improving fuel quality may not be significant relative to current retail prices, the refineries would still need to find the up-front capital costs needed to expand or upgrade processing units. Installing an isomerization unit to limit benzene and total aromatics in gasoline is estimated to cost on the order of US$20 million to US$40 million, depending on the size of the unit. Hydrotreaters for cracked naphtha (to be used as feed for reforming) are estimated to cost in the neighborhood of US$10 million–US$30 million. Marginal annual incremental operating costs vary from less than US$0.5 million to US$2.5 million.

Because refineries seldom recover the full cost of investments for improving fuel quality, it is all the more important to increase the efficiency of refinery operations in the process of implementing upgrade schemes to meet environmental regulations. Transferring ownership from the state to private operators and introducing greater competition by deregulating the downstream petroleum sector will

greatly enhance the chances of attracting the capital needed to modernize and sustain the refineries in the region.

DOWNSTREAM PETROLEUM SECTOR POLICY: DISCOURAGING MISLABELING AND SMUGGLING

The main instruments of downstream petroleum sector policy are the fixing (or not) of prices and the imposition of excise duties and taxes. Although the duties and taxes are not generally chosen with a view to encouraging the introduction of cleaner fuels, they can have a substantial influence in this direction. Governments liberalize prices to benefit from the forces of competition, and where liberalization works effectively, the domestic price is driven by movements in world product prices. Taxes then raise domestic prices above world prices plus transport costs and form an important source of government revenue.

Where prices are not market driven but are fixed by the government, they do not fluctuate regularly with world market prices but follow the world price at discrete intervals whenever the government decides to realign the domestic price. The stability that such a pricing system brings to consumers can create a disadvantage for sellers of products, who have to pay world market prices to buy product. Only when a country is completely self-sufficient in product—when it has crude and refines enough to at least satisfy its own needs—can both producers and consumers be shielded from world price fluctuations, but then the coun-

---

Box 5. Costs of Improving Fuel Quality

The incremental cost of improving fuel quality consists of additional investment costs, incremental operating and maintenance costs, and the incremental cost of importing higher-quality products. Cracked naphtha hydrotreating, reformer expansion, isomerization, and the addition of ethanol were the options considered in calculating incremental costs in this study. Linear programming models were set up for all the major refineries in the region with the exception of Turkmenbashi, which is being extensively modernized. The programs were run to examine different process and feedstock options and to estimate the incremental costs of improving the quality of transport fuel.

By running the refineries at a higher utilization rate and increasing the feed rate to the reformers, it is possible to eliminate lead in gasoline and meet vehicle octane requirements in 2005. One factor that will help to keep the incremental cost low is the slow projected growth of gasoline demand, for the reasons mentioned previously. The countries with the largest projected growth in gasoline consumption, Azerbaijan, Georgia, and Turkmenistan, do not face refinery limitations; the Nova Baku refinery in Azerbaijan and the Turkmenbashi refinery in Turkmenistan are already equipped to handle the higher gasoline demand and octane requirements, and Georgia imports almost all of its gasoline demand.

To limit benzene and total aromatics, refineries with no cracking units need to turn to isomerization, oxygenate addition, or both. Taking all the factors into account, increasing average octane to meet the vehicle fleet requirements and reformulating gasoline to eliminate lead and limit benzene to 5 vol% and aromatics to 45 vol% can be achieved at an incremental cost of about US$0.01 per liter in most countries. Retail gasoline prices in Armenia, Azerbaijan, Georgia, and Kazakhstan in 1997–99 varied between a low of US$0.17 per liter to a high of US$0.47 per liter, averaging US$0.36 per liter. Therefore, an increase of US$0.01 per liter will not prove a heavy burden for motorists.

try will suffer resource misallocation because prices are not aligned with opportunity costs.

In the region, only Azerbaijan, Turkmenistan, and Uzbekistan have the capacity to be self-sufficient in refined products. The government of Azerbaijan has fixed retail prices at constant levels for long periods. However, when domestic prices (including taxes) are much higher than potential import prices, there is an incentive to import rather than to use domestically refined products. Excise and import duties have been set at high levels to discourage legal imports, but this strategy has made illegal imports that much more attractive.

Various illegal practices in the selling of fuels are seen in the region, including smuggling, mislabeling of fuel quality, and adulteration of fuels. All are used to increase profit margins at some point in the supply chain. Often, different practices are combined; for example, a low-octane gasoline may be smuggled into a country, thus avoiding excise duty, and then sold as a higher-octane gasoline, further increasing the profit margin.

The incentive to engage in these practices depends on the potential profits, the probability of being detected, and the cost of being detected. The more efficient are customs procedures for checking the volume and quality of all imports, and the more effective is the monitoring of retail fuel quality, the less these illegal ploys will be used. High excise duties, commonly employed in the region as a fairly reliable revenue source for the government, create a strong incentive for smuggling. This is intensified by import duties (which are, however, being gradually phased out in the region). When sales are low, as they have been following the general rise in world and regional product prices, sellers have an incentive to compensate for the reduction in volume by increasing their margins through illegal practices.

Mislabeling is the simplest method for the supplier, since nothing needs to be done to the product. There is an important distinction between mislabeling for octane and mislabeling for the presence of lead. The former is more likely to be noticeable by the consumer, as the price is higher and vehicle performance is lower than the label would indicate. Mislabeling leaded gasoline is much harder for the individual to monitor, and for drivers without catalytic converters, there is little loss to using leaded; the effects are primarily external to the user.

The structure of the retail sector affects the government's ability to monitor these practices. The larger the number of independent retailers, the more difficult it is to combat adulteration and mislabeling. Where there are

large retail chains, fining one outlet can create adverse publicity for all the outlets owned by that firm, thus producing an externality to the testing. Similarly, vertical integration via franchises between distributors (who tend to be fewer in number) and retailers can spur producer self-monitoring.

As the demand for higher-quality fuels rises, partly as a result of the increase in the number of new automobiles, the market will respond by rewarding retailers who earn a reputation for supplying reliable products. At the same time the demand for testing will create commercial opportunities for the emergence of laboratories. The market for jet fuel is an example. Airlines with modern fleets will refuel in a country only if the quality of the fuel can be guaranteed. Since jet fuel is a profitable market, some local suppliers are already installing or arranging appropriate testing facilities so that they can take advantage of this opportunity.

Countries that are effectively entirely dependent on imported products will wish to buy from the most cost-effective source. There is already a substantial flow of products, both legal and illegal, from the Russian Federation. As the Russian Federation moves to more stringent fuel specifications, and particularly as lead is phased out in Russia, there is a danger that some of its refineries, as well as the marketers of gasoline lead additives, will use Central Asia and the Caucasus as an outlet for products they can no longer sell legally. Domestic duties and taxes will merely encourage further smuggling by stimulating attempts to avoid paying the duty altogether. Since excise and import duties are collected at the point of entry to the country, the higher the duty and tax element, the greater the incentive to smuggle products.

The nature of the fuels market interacts with these tendencies. The ability of users to substitute between fuels (either on purpose or by accident) creates further scope for illegal practices, some of which have an important bearing on efforts to improve fuel quality. Where fuels are very similar in their efficiency to the user, as is the case for unleaded and leaded gasoline of similar octane, attempts to discourage the use of leaded fuel by taxing it more heavily simply create an incentive to label the fuel as unleaded, since the production or import cost of leaded is slightly lower than for unleaded. It is notable that Georgia phased out such differences in tax rates prior to banning lead altogether.

Where fuels are moderately substitutable, as with gasolines of different octanes, there is an incentive to pass off a lower-octane fuel as having a higher octane if the mislabeling can escape detection. The greater the price difference, the larger the incentive.

Substitution between diesel and gasoline can take place only on the replacement of the vehicle by one using the other fuel. The difference in capital costs of vehicles (which have to be paid up-front) will dominate any savings in fuel costs per kilometer (which are spread over the lifetime of the vehicle) unless the differences in fuel prices are considerable. Here, taxation policies can play a significant role. The tax on gasoline is typically higher than that on diesel, and this tends to move demand away from leaded gasoline, but at the environmental cost of increasing emissions of the pollutants associated with diesel.

RECOMMENDATIONS FOR IMPROVING FUTURE FUEL QUALITY

In tightening transport fuel specifications in the region, the priorities are lead elimination and control of benzene emissions. Table 2 summarizes the fuel specifications in force in the region today for easy comparison with the recommended specifications given in Table 4, below. There are some country-to-country variations. For example, Armenia

limited the amount of lead in gasoline to 0.15 g/l effective March 2000, and Georgia banned leaded gasoline effective January 2000. A relatively high level of lead is permitted in high-octane gasoline.

For comparison, Table 3 lists some of the fuel parameters in the EU for two time periods, up to the end of 1999 and beginning in January 2000. The EU restricted lead in gasoline to 0.15 g/l, and has banned it altogether starting in 2000. Benzene was limited to 5 vol%, and sulfur to 0.05 wt% until last year. Effective January 2000, fuel specifications were tightened considerably; sulfur in gasoline was reduced more than threefold and benzene fivefold, and a limit on total aromatics was introduced for the first time. The limit on sulfur in diesel was lowered by 30 percent.

Taking into account health considerations, air quality, and the structure of the downstream petroleum sector in Central Asia and the Caucasus region, as well as trends in the surrounding countries, this study recommends the fuel specifications summarized in Table 4.

Complete elimination of lead by 2005 is recommended. The benzene specification is not more than 5

TABLE 2. CURRENT GASOLINE AND DIESEL SPECIFICATIONS, MAXIMUM LIMIT

Fuel	Grade	Parameter	Limits	Comments
Gasoline	76 leaded	Lead	0.17 g/l	
	93 leaded	Lead	0.37 g/l	Limited to 0.15 g/l in Armenia effective March 2000.
	All unleaded	Lead	0.013 g/l	Only unleaded gasoline is allowed in cities with populations greater than 1 million.
	All	Sulfur	0.1 wt%	
	All	Benzene	—	No limits
	All	Aromatics	—	No limits
Diesel	Type I	Sulfur	0.2 wt%	
	Type II	Sulfur	0.5 wt%	

TABLE 3. EUROPEAN UNION SPECIFICATIONS, MAXIMUM LIMIT

Fuel	Grade	Parameter	To end—1999	2000
Gasoline	Unleaded	Lead	0.013 g/l	0.013 g/l
	Leaded	Lead	0.15 g/l	banned
	All	Sulfur	0.05 wt%	0.015 wt%
	All	Benzene	5 vol%	1 vol%
	All	Aromatics	—	42 vol%
Diesel	Vehicle grade	Sulfur	0.05 wt%	0.035 wt%

TABLE 4. PROPOSED GASOLINE AND DIESEL SPECIFICATIONS, MAXIMUM LIMIT

Fuel	Grade	Parameter	2005	2015
Gasoline	All	Lead	0.013 g/l	0.013 g/l
	All	Benzene	5 vol%	2 vol%
	All	Sulfur	No change	0.03 wt%
	A76/80	Aromatics	No limit	35 vol%
	A91/93/95	Aromatics	No limit	45 vol%
Diesel	Vehicle grade	Sulfur	0.2 wt%	0.05 wt%

Note: The timing and the compositional limits for 2015 should be reassessed in a few years' time.

vol% in 2005 and 2 vol% in 2015 (or earlier). Reducing benzene further to 1 vol% in line with the current EU benzene specification would likely require benzene saturation. The gasoline sulfur limit of 0.03 wt% in 2015 should ensure efficient operation of catalytic converters, which can be and should be installed once lead is completely eliminated. Total aromatics in 2015 are limited to 35 vol% in low-octane gasoline and 45 vol% in high-octane gasoline. The extended deadline gives refineries time to adjust their operations and move away from sole reliance on reformers to other process options. Targeting these benzene, aromatics, and sulfur limits for the same year would help optimize refinery investment plans.

The diesel sulfur specification of 0.05 wt%, targeted for 2015, is identical to that introduced in the United States in 1993 and in the EU in 1996. This limit was mandated in Europe and North America to reduce sulfate-based particulate emissions from diesel. The immediate objective in Central Asia and the Caucasus region should be to lower the diesel sulfur content to the Type I specification currently in force (0.2 wt%). The recommended date for implementation of the standard is 2005.

Countries with severe air pollution problems may consider introducing more stringent specifications than are recommended here. One option, mentioned earlier, is to tighten the RVP limit during the ozone season.

Conclusions and Recommendations

This study has reviewed the current status of three principal areas closely linked to urban air quality management—air quality monitoring, vehicle emissions abatement, and fuel quality—with a view to making recommendations for the coming decade. The results of data collection and analysis have underscored the need, widely acknowledged by specialists and government officials in the region, for more extensive and systematic compilation of statistics on ambient pollutant concentrations, vehicle inventory, vehicle emissions test results, fuel consumption (including octane grade), and the quality of the fuels available on the market. Such data would make possible a more rigorous assessment of current and future air quality management policies by identifying pollutants of concern, major sources of pollution, and likely future sources of pollution and by enabling cost-benefit analysis of options for mitigating air pollution. Providing reliable statistics on vehicles and fuel consumption is an important role for the government because this information facilitates the pursuit of rational investment strategies by the government and the private sector.

The study has shown that it is possible to remove lead from gasoline, increase average octane, and limit benzene in the region by 2005 at a small cost to consumers, on the order of US$0.01 per liter. The study findings stress the importance of controlling benzene and total aromatics in gasoline as lead is phased out while, simultaneously, the overall octane requirement is increasing. It is important to emphasize that merely banning the use of lead in gasoline and taking other measures does not necessarily protect public health if there is no effective mechanism for enforcing the new standards. Simultaneous adoption of the standards throughout the region would lessen the chances that products of inferior quality will be smuggled into any one country and will facilitate control of fuel quality.

The illegal addition of benzene or aromatics to gasoline is much rarer than the adulteration of gasoline with kerosene or the addition of lead and other heavy metals to gasoline. For benzene and aromatics, quality control at the refining level should suffice, for the most part. To combat adulteration with kerosene and lead, governments in the region have to strengthen their fuel quality monitoring systems.

The very rapid growth of diesel consumption over the next 10 years will substantially increase ambient concentrations of fine particulate matter and NO_x. Because NO_x is an ozone precursor, ozone levels could also increase, depending on meteorological and other conditions. Diesel emissions are likely to be affected much more by the mechanical condition of vehicles than by the quality of diesel fuel, underscoring the importance of an effective vehicle I/M program.

If current and future standards are to be enforced more effectively, equipment upgrades and staff training will be required. It may not be possible or desirable for the government to manage every aspect of monitoring and enforcement. The role of the private sector, including NGOs, should be carefully examined. Within the government, close coordination among the ministries of environment, transport, and energy, as well as with the customs and excise services, the police, and other agencies, is important for implementation of a successful air quality management strategy.

The World Bank Group may be able to assist in the implementation of some of the recommended measures by pilot testing improved monitoring systems (for fuel quality, vehicle emissions, or air quality) in one or two countries through technical assistance and learning and innovation loans (LIL). Once the program is demonstrated to be successful, it may be replicated in other countries in the region. If private sector agencies are to be involved in some of the activities, the International Finance Corporation (IFC) of the World Bank Group could participate by providing loans. If examination of refinery economics shows that a small incremental investment project for reconfiguration of refineries to produce cleaner transport fuels is commercially viable, such a project might also involve the IFC.

REFERENCES AND SELECTED BIBLIOGRAPHY

Auto/Oil Air Quality Improvement Research Program. 1997. "Program Final Report, January 1997."

Hughes, Gordon, and Magda Lovei, 1999. *Economic Reform and Environmental Performance in Transition Economies.* Technical Paper 446. Washington D.C.: World Bank.

Kojima, Masami, and Eleodoro Mayorga-Alba, 1998. "Cleaner Transportation Fuels for Air Quality Management." Energy Issues No. 13 (July). World Bank, Washington, D.C.

Lovei, Magda, ed. 1997 *Phasing Out Lead from Gasoline in Central and Eastern Europe. Health Issues, Feasibility, and Policies.* World Bank: Washington D.C. [Available in Russian in draft form]

Lovei, Magda. 1998. *Phasing Out Lead from Gasoline. Worldwide Experience and Policy Implications.* Technical Paper 307. Washington D.C.: World Bank.

National Commitment Building Program to Phase Out Lead from Gasoline in Azerbaijan, Kazakhstan, and Uzbekistan. Document ARH. CONF/BD.45. Prepared for the Fourth Ministerial Conference, Environment for Europe, Århus, Denmark, June 23–25.

Pyadichev, E. 1999. Personal communication.

USTDA (U.S. Trade and Development Agency). 1997. "Technical, Economic, and Environmental Impact Study of Converting Uzbekistan Transportation Fleets to Natural Gas Operation." Washington, D.C.

World Bank. 1997. "Elimination of Lead in Gasoline in Latin America and the Caribbean: Status Report, December 1997." ESMAP Report No. 200/97EN. Washington, D.C.

_____ 1998. "Harmonization of Fuels Specifications in Latin America and the Caribbean." ESMAP Report No. 203/98EN. June. Washington, D.C.

_____ 2000a. "Air Quality Monitoring in Central Asia and the Caucasus: Report for the Regional Study on Cleaner Transportation Fuels for Urban Air Quality Improvement in Central Asia and the Caucasus." Washington, D.C. [Available in Russian]

_____ 2000b. "Vehicle Fleet Characterization in Central Asia and the Caucasus: Report for the Regional Study on Cleaner Transportation Fuels for Urban Air Quality Improvement in Central Asia and the Caucasus." Washington, D.C. [Available in Russian]

_____ 2000c. "Downstream Petroleum Sector Analysis in Central Asia and the Caucasus: Report for the Regional Study on Cleaner Transportation Fuels for Urban Air Quality Improvement in Central Asia and the Caucasus." Washington, D.C. [Available in Russian]

ANNEXES

ANNEX A. STUDY DESCRIPTION AND RELEVANCE

The World Bank's program Cleaner Transportation Fuels for Urban Air Quality Improvement in Central Asia and the Caucasus draws together stakeholders from the environment, transport, and oil sectors in Armenia, Azerbaijan, Georgia, Kazakhstan, the Kyrgyz Republic, Tajikistan, Turkmenistan, and Uzbekistan. The environment, transport, and petroleum ministries in all eight countries have participated in the program. Canadian and British consultants, in collaboration with local specialists, have been involved in carrying out the analysis.

The program was launched in June 1999 at a workshop in Tbilisi, Georgia. Another regional workshop focusing specifically on the refining industry was held in Ashgabat, Turkmenistan, in November 1999. A final meeting pulling together all the findings and recommendations is planned for late 2000. The workshop programs are given at the end of this annex.

Three studies were commissioned as part of this program:
- Air quality monitoring in Baku, Azerbaijan, and Tashkent, Uzbekistan (AEA Technology and Hydrometeorology)
- Vehicle fleet analysis, including fuel consumption projections (Environment Canada and various local specialists)
- Downstream petroleum sector analysis (SNC-Lavalin*Comcept Canada and various local specialists).

The final reports are being prepared in Russian and will be distributed to all the stakeholders.

Regional Workshop Programs
Regional Kick-off Workshop
10–11 June 1999, Tbilisi, Georgia

10 June 1999

Topic	Speaker

Introduction
  Introduction of participants and Workshop agenda
	Givi Kalandadze
	  Ministry of Environment, Georgia

Opening remarks
  Welcoming address and remarks on the Workshop,
  Clean Fuel Study and its importance
	Zurab Tavartkiladze
	  First Deputy Minister of Environment, Georgia

Almaty Resolution revisited
  Recollection of the Almaty Resolution and Lead Phaseout
  Study as the basis for the Clean Fuel Study
	Givi Kalandadze
	  Ministry of Environment, Georgia

Urban air quality management
  Problems of urban air pollution, policy approaches and
  priorities. Issues of the phaseout of leaded gasoline and
  beyond. Need for regional cooperation
	Magda Lovei
	  The World Bank

Agenda and objectives of the Workshop
  Workshop objectives and expectations in connection with
  the Clean Fuel Study
	Martin Fodor
	  The World Bank

International trends in fuel specifications: Links with vehicle
  emissions and air quality
  Recent developments in fuel specifications, links between
  fuel quality, vehicle emissions, and air quality
	Masami Kojima
	  The World Bank

10 June 1999 (continued)

Topic	Speaker
Urban air quality monitoring system in the Region— Example of Uzbekistan Status of monitoring system, issues and problems	Farhad Sabirov State Committee for Nature Protection, Uzbekistan
Vehicles and urban air quality Vehicle efficiency, fleet structure, vehicle emissions, fuel requirements, approaches to emission reduction	Greg Rideout and Jacek Rostkowski Environment Canada
Vehicle fleet in the Region—Example of Georgia Vehicle fleet in Georgia, recent developments, future trends and implications for fuel requirements	Elizabari Darchiashvili Ministry of Transport, Georgia

11 June 1999

Topic	Speaker
Regional programs for lead phaseout Experience of bilateral agencies with regional lead phaseout programs	Ulla Bendtsen Danish Environmental Protection Agency
Fuel quality improvement in Latin America and the Caribbean Experience and lessons learned in phasing out leaded gasoline, and overall regional program	Masami Kojima The World Bank
Lead phaseout in the Region Update on the status and ongoing activities of lead phaseout in the Lead Phaseout Study: progress, issues, next steps	Rauf Muradov NEAP Coordinator, Azerbaijan Mambet Malimbaev Lead Phaseout Coordinator, Kazakhstan Kobilzhon Abdukhamitovtch Ibragimov Uzbekistan
UNDP Lead Phaseout Project in Georgia	Givi Kalandadze Ministry of Environment, Georgia
Fuel standards in the Region Status of standardization of fuels in the region, function of the Mezhdugosudarstvennyj Komitet and national standardization bodies	Vladimir Bulatnikov Mezhdugosudarstven-nyj Komitet po Standardizacci, Russia
Regional issues in fuel quality Lead phaseout and cleaner fuel related activities and plans in the countries of the Region	Representatives from Armenia, the Kyrgyz Republic, and Tajikistan
Cleaner fuels for clean air in the Region Detailed steps for the implementation of the regional study, information and data needs, institutional support, cooperation, and organizational issues, future workshops, involvement of consultants, next steps	Magda Lovei, Masami Kojima, and Martin Fodor The World Bank
Conclusion	Givi Kalandadze Ministry of Environment, Georgia

Regional Refining Sector Workshop
4–5 November 1999, Ashgabat, Turkmenistan

4 November 1999

Topic	Speaker
Opening address	Khoshgeldy Babayev Deputy Minister of Oil and Gas, Turkmenistan
Introduction to the regional Cleaner Fuels Study, workshop agenda and organization	Martin Fodor The World Bank
Progress of the regional Clean Fuels Study to date	Martin Fodor The World Bank
Regional vehicle fleet fuel requirement	Jacek Rostkowski Environment Canada
Processes and upgrades for fuel quality improvement: coker naphtha upgrading, reforming and isomerization, and ethanol blending	John Clark SNC Lavalin*Comcept
Analysis of refinery industry in Azerbaijan, Kazakhstan, and Uzbekistan: preliminary findings	John Clark SNC Lavalin*Comcept
Panel discussion Regional perspective of the fuel quality improvement	Refinery Representatives: Novobaku, Azerbaijan; Atyrau, Pavlodar, and Shymkent, Kazakhstan; Fergana and Bukhara, Uzbekistan
Concluding remarks and Friday's agenda	Martin Fodor The World Bank

5 November 1999

Topic	Speaker
Slovak experience with lead phaseout and fuel quality improvement	Daniel Bratsky Dusan Stacho, Slovnaft
Hungarian experience with lead phaseout and fuel quality improvement	Katona Antal Danube Refinery
Experience of the European Bank for Reconstruction and Development in the region	Jaap Sprey European Bank for Reconstruction and Development
Fuel quality monitoring in Azerbaijan and Georgia	John Clark SNC Lavalin*Comcept
Fuel parameter issues in the region	Representatives from Azerbaijan and Georgia
Conclusions and closing remarks Next steps in the Clean Fuels Study	Martin Fodor The World Bank

ANNEX B. HEALTH IMPACTS OF VEHICULAR EMISSIONS

International experience indicates that, typically, *lead* and *fine particulate matter* pose the greatest health concerns in the urban environment. Lead, used to enhance octane, is one of the highest-risk pollutants still widely in use in gasoline in Central Asia and the Caucasus. The consequences of exposure to lead include diminished intelligence, behavioral and learning problems, lower productivity, and increased health care costs. Lead is especially harmful to the developing brain and the nervous system of small children, who retain significantly more of the lead to which they are exposed than do adults. Lead poisoning affects the poor disproportionately; the available data suggest that more lead is absorbed in a situation of iron or calcium deficiency, and the amount of lead absorbed by the body increases significantly when the stomach is empty. Unlike other pollutants, lead does not degrade; it continues to accumulate in the environment until its use is stopped.

After lead, exposure to fine particulate matter typically causes the most serious health damage in urban areas. Particulate matter smaller than 10 microns (PM_{10}), and especially particulate matter smaller than 2.5 microns ($PM_{2.5}$), are responsible for high incidences of respiratory infections, resulting in lost work days, hospitalizations, and premature death. Coarse particles are believed to have a less significant effect on health, and measures of dust or total suspended particles (TSP) are therefore less relevant than measures of fine particles. Particulate emissions from combustion, including the combustion of liquid fuels in vehicles, fall predominantly in the $PM_{2.5}$ range. Diesel emissions are a significant source of such emissions. This is of special concern because demand for diesel is increasing rapidly throughout Central Asia and the Caucasus as heavy-duty vehicles switch from gasoline to diesel to take advantage of higher fuel economy and greater engine durability. Furthermore, there is mounting epidemiological evidence that diesel exhaust poses a serious cancer risk.

All combustion and metallurgical processes and many other industrial operations lead to the emission of particles into the atmosphere. Particles emitted directly from a source are termed *primary*. Particles formed within the atmosphere, mostly from the chemical oxidation of atmospheric gases, are termed *secondary*. Oxides of sulfur and nitrogen are among the contributors to formation of secondary particles.

Other pollutants linked to the transport sector that have public health impacts include carbon monoxide (CO), oxides of nitrogen (NO_x), oxides of sulfur (SO_x), ozone, and airborne toxics.

- CO, a colorless, odorless gas that inhibits the capacity of blood to carry oxygen to organs and tissues, is a product of incomplete combustion of fossil fuels. Gasoline-fueled vehicles account for the bulk of CO emissions in most cities. People with chronic heart disease may experience chest pains when CO levels are high.

- SO_x is a product of the combustion of sulfur-containing fossil fuels. Periodic episodes of high concentrations of sulfur dioxide (SO_2) cause reduced lung function in asthmatics and exacerbation of respiratory symptoms in sensitive individuals. SO_2 contributes to the formation of secondary particulate matter as it reacts with other substances in the atmosphere to form sulfate aerosols that make up a part of $PM_{2.5}$. SO_2 travels long distances and contributes to acid rain and damage to vegetation (including forests and agricultural crops) and to freshwater lakes.

- NO_x contributes to acid rain and secondary particulate formation and it is a precursor for ground-level ozone. Nitrogen dioxide (NO_2), emitted by both gasoline and diesel vehicles, causes changes in lung function in asthmatics.

- Ozone is formed by reactions with NO_x of photochemically reactive organic compounds, commonly referred to as VOCs (volatile organic compounds, including aldehydes, olefins, and aromatics with two or more alkyl groups, which are ozone precursors). Ozone is responsible for photochemical smog and has been associated with transient effects on the human respiratory system. Of the documented health effects, the most significant is decrements in the pulmonary function of individuals participating in light to heavy exercise. As income levels rise, the ownership of passenger cars increases and, depending on meteorological conditions, ozone can become a serious problem.

- Toxic emissions from vehicles include benzene, polycyclic aromatics, 1,3-butadiene (a potent carcinogen), and aldehydes. The WHO lists acetaldehyde, benzene, diesel exhaust, and polycyclic aromatics as carcinogens and provides guidelines for ambient concentrations.

ANNEX C. AIR QUALITY MONITORING IN BAKU, AZERBAIJAN, AND TASHKENT, UZBEKISTAN

DATA COLLECTED USING AUTOMATIC SAMPLERS IN BAKU, AZERBAIJAN

Data were collected in Baku at one of the observation posts operated by the national Hydrometeorology Department. Monitoring at a position colocated with the local monitoring equipment enabled comparison of the data obtained using the two techniques. The observation post was located in an "uptown" region of Baku, approximately 5 meters from curbside and 15 meters from the center of the road. The data collected are presented in Figure C-1 and summarized in Table C-1.

From the plot of the data, it can be seen that levels of all pollutants vary considerably during the study—a trend that cannot be captured by the current monitoring system used by the Hydrometeorology Department.

Analysis of the data indicates that WHO health guidelines were exceeded on two occasions, for NO_2 when the maximum hourly average reached 212 micrograms per cubic meter ($\mu g/m^3$), the guidance value being 200 $\mu g/m^3$. WHO guidance values were not exceeded for any of the other pollutants during the period of the survey.

TABLE C-1. SUMMARY STATISTICS, BAKU

	NO_2 ($\mu g/m^3$)	CO (mg/m^3)	O_3 ($\mu g/m^3$)	TSP (dust) ($\mu g/m^3$)
Average	88	2	73	38
Hourly maximum	212	5.2	95	124
Eight-hour maximum	176	4.2	88	94
Daily maximum	121	2.9	83	64

Note: $\mu g/m^3$, micrograms per cubic meter; mg/m^3, milligrams per cubic meter.

FIGURE C-1. AIR QUALITY DATA, BAKU, AZERBAIJAN

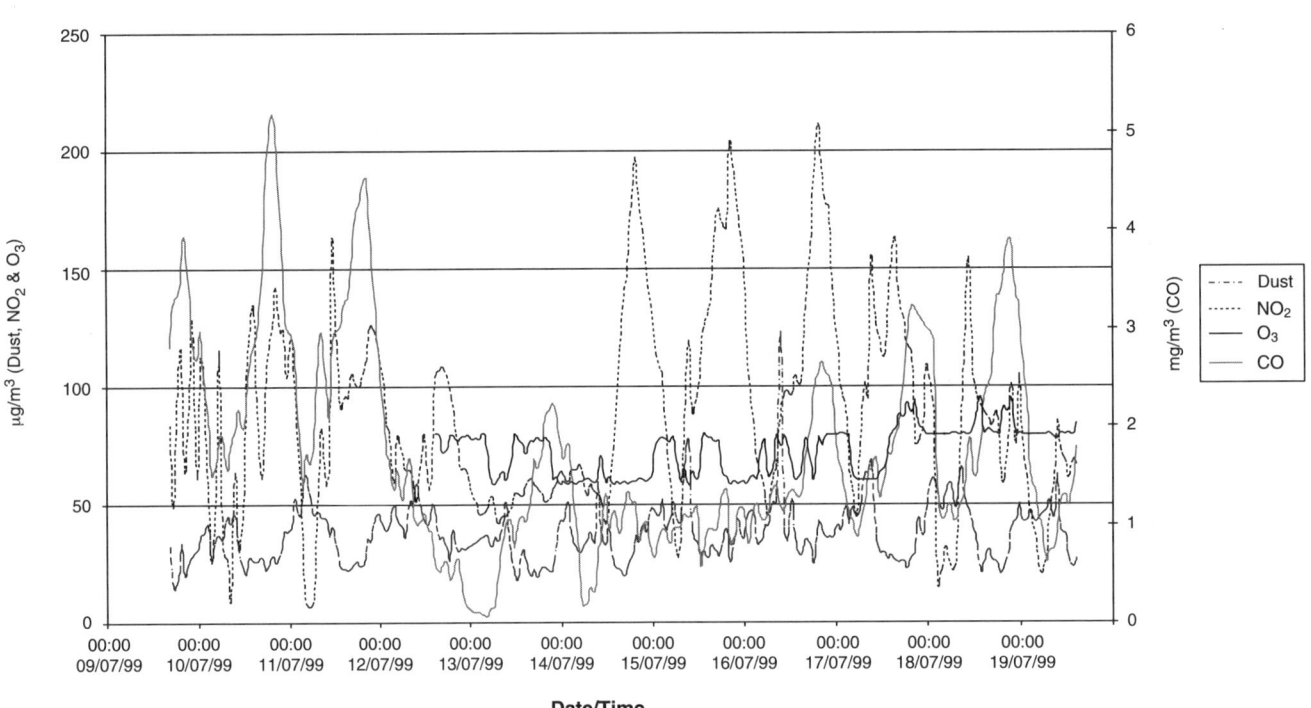

DATA COLLECTED USING AUTOMATIC SAMPLERS IN TASHKENT, UZBEKISTAN

As with the monitoring in Baku, the automatic analyzers in Tashkent were colocated with a Hydrometeorology Department observation post, chosen as the one likely to give the highest values for the pollutants concerned during the monitoring period. The post was situated on a major crossroads in the center of the city. Data are not available for TSP because the analyzer used to collect data for that pollutant was damaged in transit to Uzbekistan. In addition, as a result of frequent interruptions in the power supply to the monitoring station, data capture (the number of valid data points) for the other pollutants was appreciably lower than in the Baku survey. Where, because of power outages, the analyzer was not functioning, the corresponding data points are ignored for the purposes of calculating averages and statistics. The data collected are presented in Figure C-2. Table C-2 lists summary statistics. As with the data from Baku, the plot of the data from Tashkent shows that the levels of all pollutants varied considerably during the study.

Analysis of the data indicates that WHO health guidelines were exceeded on 7 occasions for NO_2 and on 15 occasions for ozone. The maximum recorded hourly average was 250 $\mu g/m^3$ (the guidance value is 200 $\mu g/m^3$). The ozone exceedances were against the running eight-hour mean guidance value of 120 $\mu g/m^3$; the maximum recorded eight-hour average was 162 $\mu g/m^3$.

Guidelines were not exceeded for CO during the period of the survey, although the running eight-hour mean limit value of 10 milligrams per cubic meter (mg/m^3) was approached on several occasions, with the maximum value being 8.7 mg/m^3.

TABLE C-2. SUMMARY STATISTICS, TASHKENT

	NO_2 ($\mu g/m^3$)	CO (mg/m^3)	O_3 ($\mu g/m^3$)
Average	90	4.2	72
Hourly maximum	250	13.9	218
Eight-hour maximum	171	8.7	162
Daily maximum	109	5.7	75

FIGURE C-2. AIR QUALITY DATA, TASHKENT, UZBEKISTAN

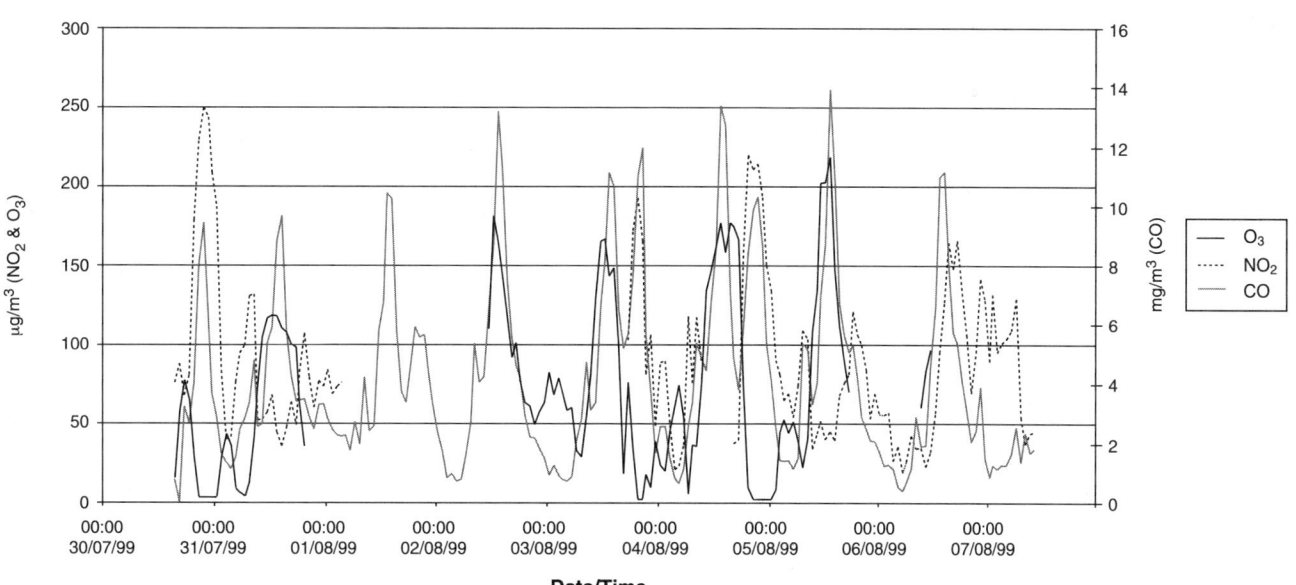

COMPARISON OF DATA

Figures C-3 through C-5 compare the data obtained by AEA Technology using automatic analyzers with those obtained by the local Hydrometeorology Department during the same period at the same observation post.

Comparison of the NO_2 values reported by Hydrometeorology and AEA show that agreement ranges from very good (July 10) to a difference of a factor of two (July 15). Comparisons of CO and TSP concentrations are complicated by the fact that the results reported by Hydrometeorology are at the detection limits for the techniques used, whereas the instruments used by AEA are much more sensitive to lower concentrations.

FIGURE C-3. COMPARISON OF DAILY NO_2 DATA, BAKU

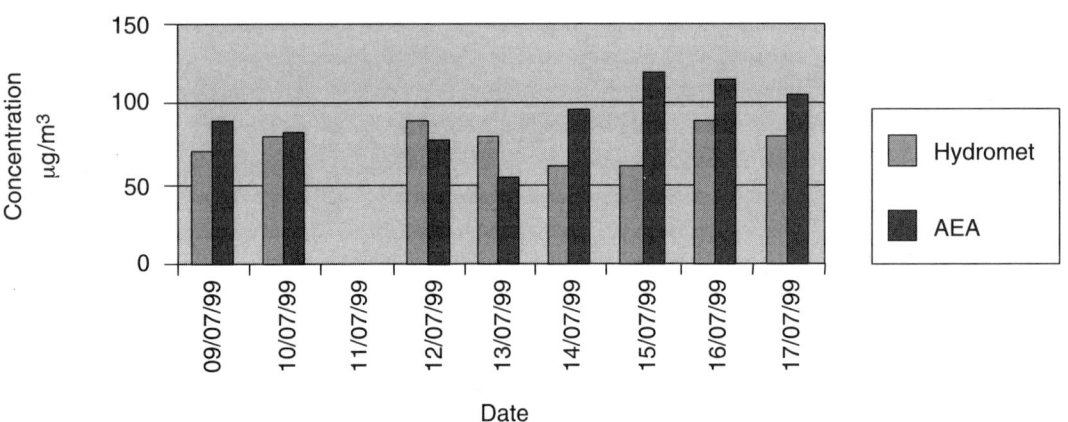

FIGURE C-4. COMPARISON OF DAILY CO DATA, BAKU

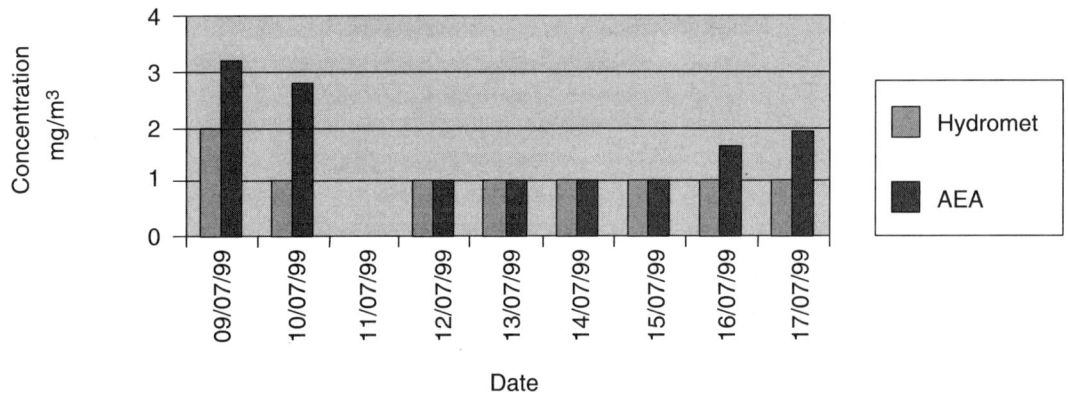

FIGURE C-5. COMPARISON OF DAILY TSP DATA, BAKU

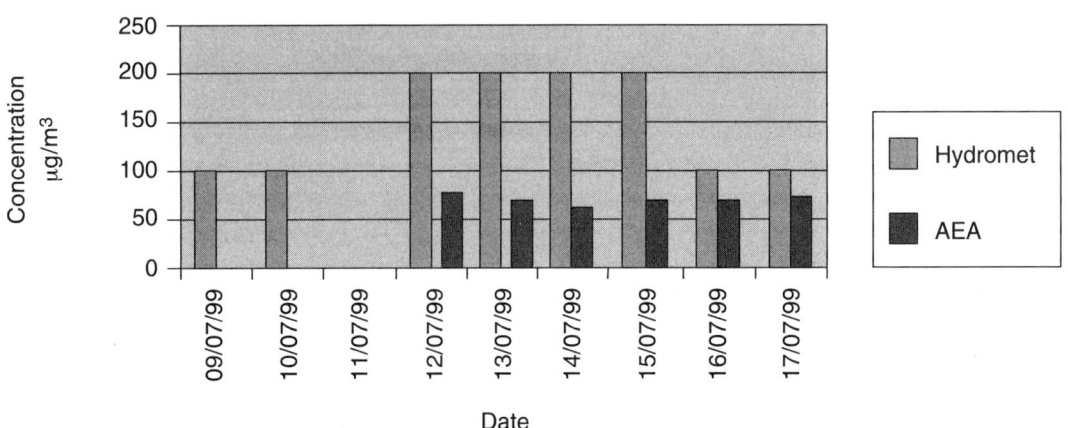

ANNEX D. MOTOR FUEL QUALITY AND REFORMULATION

This annex reviews considerations that should be taken into account when considering fuel quality and key principles in developing fuel specifications.

Lead has historically been added to gasoline as an octane enhancer to ensure smooth combustion. Eliminating lead in gasoline is the first step in improving motor fuel quality. Experience elsewhere has clearly demonstrated that elimination of gasoline lead should be carried out within the broader context of improving fuel quality and introducing an integrated approach to air pollution management. This is because gasoline components that are added to compensate for the octane shortfall after lead removal can have harmful effects of their own. There are concerns about increased emissions of carcinogens such as benzene (from higher-severity reformer operation) and polycyclic aromatics (from catalytic crackers). Another potential problem is ozone precursors such as olefins (from catalytic crackers), di- and tri-alkylaromatics (from higher-severity reformer operation), and NO_x (from combustion of gasoline containing higher levels of aromatics and olefins, particularly in the absence of catalytic converters). Because of the prevalence of thermal inversion and sunny weather, ozone may become a serious problem in many cities in the region once the volume of traffic starts to increase significantly. Photochemical smog has already been observed in some cities, including Tashkent. In modernizing refineries to phase lead out of gasoline, it will be cost-effective to address these issues in gasoline reformulation at the same time, thereby optimizing refinery investment plans.

Benzene is a growing focus of attention worldwide. Gasoline vehicles are primarily responsible for benzene emissions, which in turn arise from two sources: evaporative and exhaust emissions of benzene in gasoline, and exhaust emissions of benzene from incomplete combustion of nonbenzene aromatics in gasoline. Benzene in gasoline contributes 10 to 20 times more to exhaust benzene emissions than does incomplete combustion of nonbenzene aromatics. Catalytic converters are effective in reducing exhaust benzene emissions. From the point of view of refinery economics, it is generally more cost-effective to limit benzene itself than to limit total aromatics. Limitation of total aromatics, which are formed in a reformer, is costly to the refiner because a by-product of aromatics production is hydrogen, which is needed to reduce sulfur in gasoline and diesel. Reformers are the cheapest source of hydrogen.

Aromatics with two or more alkyl branches are ozone precursors. A U.S. automobile and oil industry study found, however, that for cars equipped with catalytic converters, reducing aromatics from 45 to 20 percent had no impact on ground-level ozone (Auto/Oil Air Quality Improvement Research Program 1997). A cost-effective way of controlling ozone precursors, which include olefins, is to limit gasoline volatility. Reducing light olefins [as a result of lowering Reid vapor pressure (RVP)] has a significant effect on formation of ground-level ozone.

Oxygenates enable more complete combustion of gasoline, resulting in lower exhaust CO and hydrocarbon emissions from cars not equipped with oxygen sensors. They also act as a diluent, lowering the amount of aromatics, olefins, and benzene in gasoline. Oxygenates are extremely high in octane, easing lead removal. Ethers and alcohols have been added to gasoline for this reason since the 1970s. MTBE, the most common ether added to gasoline, has a reasonably low RVP and high octane, and typically up to 15 vol% is added. A disadvantage of oxygenates is that they are miscible with water. Recently, there has been concern in the United States after studies showed that leaking underground storage tanks have led to contamination of groundwater with MTBE. The levels detected are typically much lower than the range of exposure levels at which health effects are observed in animal toxicology studies. MTBE, however, has a very low odor and taste threshold; it is detected at a level much below the health threshold level, and consumers complain. Ethanol is an alternative, but it is expensive in North America and has required government subsidies. Ethanol at low concentrations has a high blending RVP, so that adding it makes it more difficult to use such cheap sources of octane as butane.

One of the most effective ways of lowering exhaust emissions is the use of catalytic converters. Catalysts, however, can be poisoned. Lead is a permanent poison, and leaded gasoline should never be used in cars equipped with catalytic converters. Sulfur is a temporary poison; its deactivating effects are reversible to a large extent. For efficient catalyst operation, the level of sulfur in gasoline should be limited to 0.05 wt%, and preferably lower.

A potential future concern is fine particulate pollution, for which diesel is a significant source. The quality of diesel fuel in the region is reasonable for the purpose of controlling particulate emissions. Sulfur is the primary parameter that may need to be controlled in the future. Vehicle maintenance and improved engine design are expected to play a key role in minimizing diesel particulate emissions.

ABBREVIATIONS AND ACRONYMS

CEE	Central and Eastern Europe
CNG	compressed natural gas
CO	carbon monoxide
DEPA	Danish Environmental Protection Agency
EBRD	European Bank for Reconstruction and Development
EU	European Union
FCC	fluidized catalytic cracker
GDP	gross domestic product
g/l	grams per liter
IFC	International Finance Corporation (of the World Bank Group)
I/M	inspection and maintenance
kg	kilogram
km	kilometers
kPa	kilopascals
LIL	learning and innovation loan
LNG	liquefied natural gas
mg	milligram, 10^{-3} gram
mg/m^3	milligrams per cubic meter
MON	motor octane number
MTBE	methyl tertiary butyl ether
NEAP	national environmental action program
NIS	new independent states
NGO	nongovernmental organization
NO$_2$	nitrogen dioxide
NO$_x$	oxides of nitrogen
PM$_{2.5}$	particulate matter with an aerodynamic diameter smaller than 2.5 microns
PM$_{10}$	particulate matter with an aerodynamic diameter smaller than 10 microns
RON	research octane number
RVP	Reid vapor pressure
SO$_2$	sulfur dioxide
SO$_x$	oxides of sulfur
T96	temperature at which 96 percent of the fuel evaporates
TSP	total suspended particles
UNDP	United Nations Development Programme
UNECE	United Nations Economic Commission for Europe
US$	United States dollars
VKT	vehicle kilometers traveled
VOC	volatile organic compound
vol%	volume percent
WHO	World Health Organization
wt%	weight percent
μg	microgram, 10^{-6} gram
μg/m^3	micrograms per cubic meter